PYTHON PROGRAMMING FOR YOUNG CODERS

A Hands-On, Project-Based Introduction to Coding for Beginners, Kids, and Teens

Anand Pandey

SCRIB INK
PUBLISHING

Python Programming for Young Coders: A Hands-On, Project-Based Introduction to Coding for Beginners, Kids, and Teens, 1st Edition.

Copyright © 2025 by Anand Pandey

All rights reserved. No portion of this book may be reproduced in any form without written permission from the publisher or author, except as permitted by U.S. copyright law.

No part of this book may be reproduced, distributed, or transmitted in any form or by any means, including photocopying, recording, or other electronic or mechanical methods, without the prior written permission of the publisher or author, except in the case of brief quotations embodied in critical reviews and certain other noncommercial uses permitted by U.S. copyright law. For permission requests, write to the publisher at the address below.

Publisher:
Scrib Ink Publishing.
Plymouth Meeting, PA, USA
info@scrib.ink
www.scrib.ink

ISBN: 978-0-9997408-6-6

Disclaimer: While the author and publisher have made every effort to ensure the accuracy and completeness of information contained in this book, we assume no responsibility for errors, omissions, or contrary interpretations of the subject matter herein. This book is for informational purposes only and is provided on an "as-is" basis. The author and publisher shall not be liable for any losses or damages arising from the use of this book.

Some references used in this book are derived from The Python Language Reference by Guido van Rossum and others, published by the Python Software Foundation. For detailed information, please visit the official Python documentation: http://docs.python.org/py3k/reference/index.html.

This book references and discusses the Mu Editor, an open-source Python IDE created by Nicholas H. Tollervey and contributors. Mu Editor is licensed under the GNU General Public License (GPL) v3.0. For more details, visit https://github.com/mu-editor/mu.

This book references and discusses Trinket, an online Python IDE and platform for writing, running, and sharing Python code, created by the team at Trinket. Trinket is licensed under its own terms and conditions. For more details, visit https://trinket.io.

Some of the icons used in this book were designed by Freepik (https://www.freepik.com/) and are used in compliance with their licensing terms.

Table of Contents

ABOUT THE AUTHOR 9

INTRODUCTION 13

GETTING READY FOR PYTHON PROGRAMMING 15

    PROGRAMMING LANGUAGE 15
    PROGRAM VS CODE 16
    WHY LEARN CODING? 18
    WHAT IS PYTHON? 19
    WHAT CAN YOU DO WITH PYTHON? 21
    WHY IS IT NAMED PYTHON? 22
    HOW LONG DOES IT TAKE TO LEARN PYTHON? 22
    WRITING A PYTHON PROGRAM 23
        Getting Started with Mu Editor 23
    RUNNING A PYTHON PROGRAM 27
        What Is a Script? 28
    CODE FORMATTING IN THE BOOK 32
    REVIEW QUESTIONS 33

DRAWING WITH PYTHON 35

    DEGREES 38

VARIABLES 45

    VARIABLE ASSIGNMENT STATEMENT 48
    VARIABLE NAMING RULES 49
    REVIEW QUESTIONS 51

FUNCTION BASICS 53

    PYTHON'S BUILT-IN FUNCTIONS 55
        Printing Python's Output – The `print()` *Function* 55
        Receive User Feedback – The `input()` *Function* 58
    REVIEW QUESTIONS 60

DATA TYPES 63

INTEGERS	64
FLOATING POINT NUMBERS	65
STRINGS	66
BOOLEAN	68
OTHER DATA TYPES	69
DATA TYPE CONVERSION	69
Converting Strings to Integers	*70*
Converting Strings to Floats	*70*
Converting Numbers to Strings	*71*
REVIEW QUESTIONS	72

CODE EXECUTION FLOW 75

OPERATORS 83

TYPES OF OPERATORS	84
ARITHMETIC OPERATORS	84
COMPARISON OPERATORS	85
ASSIGNMENT OPERATORS	87
Chaining Assignments	*88*
LOGICAL OPERATORS	88
Logical operator precedence	*90*
REVIEW QUESTIONS	92

NUMBERS 95

MATHEMATICAL EXPRESSIONS	96
NUMBERS - INTEGERS AND FLOATS	97
CONVERTING NUMBERS	97
int()	*98*
float()	*98*
BASIC ARITHMETIC OPERATORS	99
NEW OPERATORS	102
Floor and Ceiling	*102*
Floor Division	*103*
Exponentiation	*104*
MODULO	105
RANDOM NUMBERS	106
REVIEW QUESTIONS	109

STRINGS 111

String Creation 112
String Concatenation 114
Length of a String 115
Repetition 115
Accessing Characters in a String (Indexing) 116
Slicing Strings 118
String Formatting 119
String Methods 119
Changing Case (Uppercase / Lowercase) 120
Capitalize 120
Title 121
Strip 121
Replace 122
Find 122
Review Questions 123

LIST, TUPLES, AND DICTIONARIES 125

Data Structure 126
List 127
Creating a List 129
Nested List 131
Creating a nested list 131
Accessing List Elements 132
Slicing a List 132
Adding Elements to a List 133
Creating an empty list and adding items later 134
Starting with a list and expanding it 134
Inserting an element 134
Extending a list 135
Remove Elements from a List 135
Popping an Element from a List 136
Clearing a List 137
Modifying a List 137
Finding Elements 138
Total Number of Occurrences 139

    Length of a List 139
    List Concatenation 140
  Tuple 140
    Creating Tuples 140
    Supported Methods 142
  Dictionary 142
    Creating a Dictionary 142
    Accessing Dictionary Elements 143
    Adding to Dictionary 145
    Modifying a Dictionary 145
    Removing Items from a Dictionary 146
    Length of a Dictionary 147
    Merging Dictionaries 147
    Dictionary Unpacking 148
    Nested Dictionaries 149
  Review Questions 151

CONDITIONAL STATEMENTS 153

  If Statement 154
  Else Condition 156
  Multiple Conditions (if-elif) 157
  Logical Operators in Conditionals 158
  Nested Conditionals 159
  Truthiness (Truthy Vs Falsy) 160
  Review Questions 165

LOOPS 167

  range() Function 170
  For Loop 172
  While Loop 175
  Review Questions 179

FUNCTIONS 183

  Function Definition 187
  Calling a Function 188
  Parameters and Arguments 189

Return Statement	189
Default Parameter	190
Recursive Functions	191
Documenting Functions	193
Tracing Function Call	196
Review Questions	209

MODULES 211

Types of Modules	212
Importing Modules	214
Import with Alias	*214*
Import Specific Functions or Classes	*214*
User Defined Modules	215
Use __name__ and __main__ to Avoid Running Code As a Script	*216*
Review Questions	217

OBJECT-ORIENTED PROGRAMMING 219

Class and Objects	220
Attributes	222
Methods	224
Encapsulation	225
Inheritance	228
Review Questions	231

PROJECT – LIBRARY MANAGEMENT SYSTEM 233

Requirements	234
Design Thinking	235
Test Scenarios:	*238*
System Architecture and Design	238

INDEX 248

ERRORS & UPDATES

At **Scrib Ink Publishing**, we are committed to providing accurate and high-quality content. If you come across any errors, please visit www.scrib.ink for errata and updates.

We appreciate your support and strive to keep improving your learning experience!

About the Author

Anand Pandey is a technology professional with extensive experience in the industry. He holds a master's in data science. He has used Python throughout his career to help businesses solve complex problems through technical solutions. He authored this book to share his knowledge and passion with young learners.

Beyond his professional work, Anand enjoys coding as a creative and problem-solving tool. He uses his programming skills to innovate, write code, and address everyday challenges by creating tools and utilities. He believes that coding is more than just a career—it's a way of thinking and solving problems.

Anand firmly believes that programming is an essential skill for every child. In a world increasingly shaped by technology, learning to code not only prepares children for a digital future but also develops an organized, problem-solving mindset. It encourages critical thinking, logical reasoning, and creativity, which are invaluable in navigating life's challenges. Through this book, he hopes to inspire young minds to embrace coding and develop these vital skills.

> *Coding is today's language of creativity. All our children deserve a chance to become creators instead of consumers of computer science.*

MARIA KLAWE
FORMER PRESIDENT OF HARVEY MUDD COLLEGE

Introduction

Take a moment and think of a world where you do not have apps on your phone, video games on your computer, or even the internet itself. It is hard to believe, right? Luckily, we have computer programs that enable us to create all these things. Programs are powerful. They're behind most aspects of our modern-day lives—from the games you play to how you talk to your friends online and even how doctors help people get better. While shopping, you might have seen your parents swiping their credit cards. Computer programs read that information and complete the transactions with the bank. Doesn't it sound interesting?

Once you learn about programming, you can write code to make games, build robots, and help solve some of the biggest problems in the world.

Python is one of the most popular programming languages for beginners. It does not present itself with a complex set of rules and syntax. It has an English-like syntax, which makes it a beginner-friendly language. Python is also a versatile language used for a variety of tasks and in different industries.

This book is written keeping beginners in mind. You don't need to know any programming to get started with this book. I've used real-life examples wherever I could to make the concepts easy and fun to understand. If you already know a little programming, you might go through the book faster. However, I still recommend reading every chapter to reinforce the concepts. Now that Artificial

Intelligence (AI) and Machine Learning (ML) are adopted widely, knowing Python is more important than ever.

This book takes a unique approach to learning Python. It doesn't just teach you important concepts but also helps you create something new along the way. There are plenty of books out there, but I wanted to write something different—something that teaches coding in a fun, exciting, step-by-step way. Each chapter has quizzes to help you check what you've learned so far before moving on. The answers to the quizzes are at the end of each chapter.

This book is your first step toward exploring a world of possibilities. While there is much more to coding than what is covered here, consider this book as the key that opens the door to countless opportunities. No single book can cover every concept, but my goal is to spark your curiosity and set you on the right path for continued learning.

This book isn't just about teaching you the rules of coding; it's about helping you use those skills to create something amazing. I hope this book makes learning Python as fun for you as it was for me to write it. Honestly, I wish I had a book like this when I was starting out—it would have made coding so much easier and way more exciting.

Now it's your turn. Let's dive in and start this awesome coding adventure together.

1

Getting Ready for Python Programming

Introduction to Programming and Environment Setup

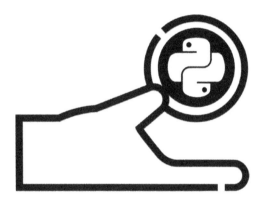

Programming Language

If you had to communicate with someone from France and did not speak their language, how would you respond? Wouldn't you learn French? The same is true if you wish to communicate with computers. Computers have their own languages too! These are called programming languages. Python is one example of a programming language. These languages have special rules and symbols that the computer understands. Programmers use these languages to give instructions to computers, just like you use words to talk to your friends. These instructions are called "code". It's like writing a secret message with special rules and symbols the computer can understand. It's like writing a story, but instead of words, you use code to tell the computer what to do. If you know how to code, you can instruct the computer to do incredible things, such as create games, make websites, control robots, create apps, and much more.

In short, coding is all about giving instructions to a computer. It's a fun and creative way to make your ideas come to life.

Program Vs Code

People often confuse "program" and "code" and use them interchangeably. But I want to shed some light on this to help build a solid foundation and properly understand these concepts. As explained above, code is a set of instructions that we write by translating our language into a machine-readable format using a specific programming language (like Python, Java, C, etc.). So, what is a program? Programming is a broader concept that is not just about providing instructions to computers but also includes the entire process of creating the code. Wait! What are these processes? Well, when you design a computer project, you do not start with writing code. You need to start with planning, designing, developing algorithms, architecture, and deployment strategy. So, programming has a broader scope, and coding is just one part of it.

Chapter 1

Let's play a quick game to understand coding better. This little girl wants to play with all her toys in the box. But she doesn't know the way! We need to help her get there. We can give her instructions, like 'Go Forward,' 'Turn Left,' or 'Turn Right.' It's like giving her a map.

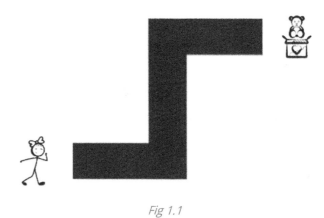

Fig 1.1

Let's see, how can we get her to the toys? Hmm... maybe we could say the following (Fig 1.2):

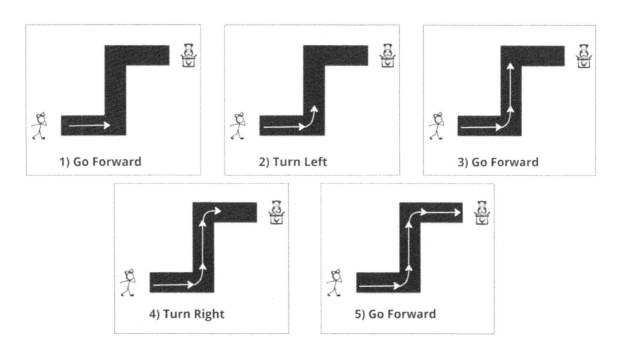

Fig 1.2

These are our 5-step instructions, or our 'code.' By following this code step-by-step, the girl can finally reach her toys.

I showed you this example to explain how we will think step-by-step to solve any coding problem. Here, we saw our problem statement in Fig. 1.1 and then immediately translated it into step-by-step instructions. These instructions can now be easily fed into a computer using a programming language like Python.

Why Learn Coding?

There are many reasons to learn how to code, but I'll focus on the most important ones. Have you ever solved a big, complicated jigsaw puzzle? At first, it looks hard to solve, right? The thought that comes to your mind is, "Whoa! How will I ever solve this? Is there an easy way out?" But then you roll up your sleeves, put your nose to the grind, and start solving it with a mix of curiosity and problem-solving strategies, which mostly include the following:

1. You gather as much information as you can.
2. You use visual skills to identify shapes, patterns, and colors to understand where each piece might fit.
3. You try problem-solving techniques, like thinking, "This shape might go here, but wait! It's slightly uneven on this edge…"
4. You experiment with trial and error.
5. You identify issues and fix them and sometimes refactor some of the steps you took earlier – maybe you placed some pieces incorrectly. But your sharp eyes catch the mistake, and you adjust the pieces to put them in the right place.

6. Finally, you test your work – you run a cursory look across the puzzle board to make sure everything is done to perfection.

This is exactly what you do when you write code to solve a real-world problem. Coding teaches you to think in a structured manner with a clear strategy and goal. It builds resilience and helps you not only write quality instructions for computers but also tackle challenges in other areas – like understanding a tricky math problem or figuring out the best way to win a game.

Coding gives you instant gratification. When you finish a coding project, you feel like a superhero! It gives you a big confidence boost, like saying, "Wow, I did that!" This feeling motivates you to tackle even harder challenges, not just in coding but in anything you want to achieve.

Coding is like having a magic wand for your imagination. How about building a game and then playing it? Wouldn't that be awesome? You can create cool games, like a space adventure or a funny animal race. You can even design incredible pictures and animations that move and dance on the screen.

So, see? Coding isn't just about learning a special language for computers. It's about becoming a better problem-solver, unlocking your creativity, and feeling super proud. It's like a superpower that helps you grow and succeed in everything you do.

What Is Python?

In the above section, we saw that Python is one example of a programming language. It's a very friendly English-like language that uses words and rules that are easy to understand. It's considered a great starting point for someone who is starting to learn coding.

There are generally two types of programming languages - compiled and interpreted. Compiled languages (like C, C++, or Java) need a special program to convert the entire code to a machine-readable format before it can be executed. You might have seen some ".exe" files on your computer. They are the output from a compiled language. Once a program is compiled, it can be stored and run any number of times directly without needing further interpretation by the computer. However, every time you make changes to your code, it needs to be recompiled. This makes the development process slower.

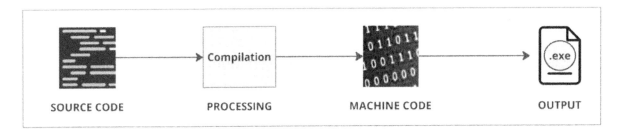

In contrast to the compiled languages, interpreted languages work differently. They do not require a separate compilation step. Python is one such language. For interpreted languages like Python, you can run the code directly, line by line, without needing a separate step for the compilation. There are several advantages of an interpreted language - the development is faster. You can make changes on the fly and re-run your code instantly. Also, if there are any errors, they will be reported as they are encountered. Interpreted languages like Python are easier for beginners as they allow quick development with immediate feedback and are easier to test and debug.

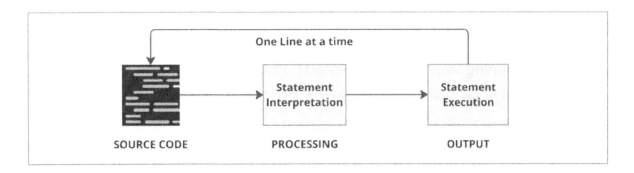

Python is widely used in thousands of real-world business applications, including numerous large-scale and mission-critical systems. Many well-known companies rely on Python for their development work. Some examples include Microsoft, Google, Amazon, Reddit, Netflix, Instagram, and NASA.

What Can You Do with Python?

Python is very popular and has a wide range of applications. Here are some of its top uses:

- Python can be used to build websites and web applications.
- Data scientists use Python extensively for tasks like data cleaning, analysis, exploring datasets, visualization, and creating insights.
- Python is used in Artificial Intelligence and machine learning to make smart applications.
- Python helps in the automation of repetitive tasks, like organizing files or sending emails, saving both time and effort.
- Python is used in game development. You can build your own games using Python.

Whether you're into games, websites, science, or robotics, Python is a great tool to help bring your ideas to life.

Why Is It Named Python?

Python programming language was created by a Dutch programmer, Guido van Rossum. The language was conceived in the late 1980s, and the implementation started in December 1989. The first full release came out in 1994. It's said that Guido van Rossum was a big fan of the British comedy series "Monty Python's Flying Circus." He named the language after this show, reflecting his fun-loving nature and desire to create something enjoyable. The playful and irreverent name aligns with Python's philosophy of simplicity and readability, making it more approachable for programmers.

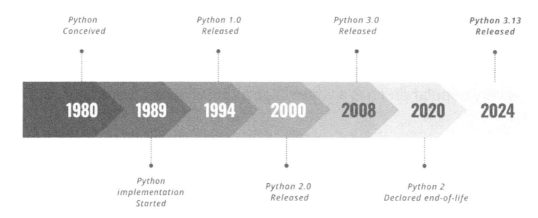

How Long Does It Take to Learn Python?

Remember when you first learned to ride a bike? Learning Python is a bit like that—it takes time, practice, and patience! How long it takes depends on your dedication, prior coding experience, and how much time you are willing to spend learning.

This book will teach you the basics of Python. It might take a few weeks, or even a few months, depending on how much time you spend on it. You can get the hang of the basics in around 2 to 3 months if you are willing to spend 5 to 10 hours each week. But here is the exciting part—you can keep learning and growing as a coder for years! Just imagine all the amazing things you will be able to create once you have mastered Python. The secret is to practice regularly and stick to it.

It is recommended that you read one chapter daily and code for a few hours each week. With steady effort, you'll see your skills improve faster than you think. Don't hesitate to ask for help if you get stuck and always remember to have fun along the way. With commitment and a dash of fun, you'll be amazed at what you can achieve.

Writing a Python Program

Before we write our first code, let's set up a place where we can run it. There are a lot of options for writing and running Python, and you might already have a favorite one. If you already have a preference, go ahead and use that. But if you're just starting, I recommend using a simple development setup using Mu Python code editor that's easy to use and gets you coding quickly. I will be using Mu Editor for the rest of the book. The Mu Editor is a beginner-friendly Python development environment designed to make programming accessible for learners. It is open-source software licensed under the GNU General Public License (GPL) v3.

Getting Started with Mu Editor

Follow the steps below to download and install Mu Editor:

1. Go to https://codewith.mu/en/download and click on the installer that works for your computer (Windows, Mac, or Linux).

2. Follow the instructions to install it. It's quick and easy.
3. Once installation is complete, open Mu by clicking on its icon.
4. Since its first time you will use Mu, it will ask you to choose a mode. Pick "Python" ("Python3" in case of macOS) and click OK.

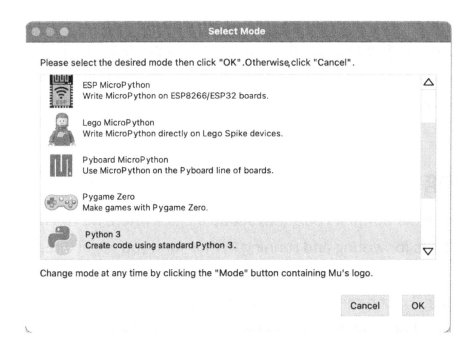

5. Now, it's time to write your first code. Paste the following code at line # 2. Click "Run" button from the top menu.

Chapter 1

Press this to run the code

6. It will now ask you to save the file. Give a name test.py and click "Save". A screen (called the console) will appear at the bottom, showing the output of your code.

Output

7. Now, Mu Editor is setup for you to start coding.

If you don't want to download and install anything on your local computer, you can use a browser-based Python editor like Trinket as an alternative. Trinket is an online platform that allows you to write, run, and share Python code directly in your web browser. To get started with Trinket, visit their website at https://trinket.io/ and click on the "Help" menu item in the top right corner of the page, as shown below:

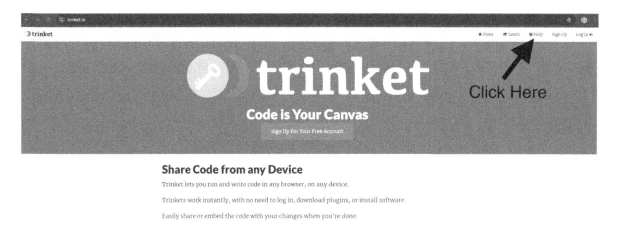

On the next screen, click the "Make your first trinket" button, as shown below:

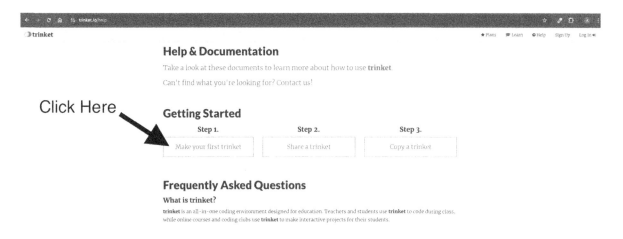

This will open a screen that guides you step-by-step through running a Python program using Trinket online. Just remember to select Python 3 when prompted to choose a version.

Your final, code-ready screen should look like this:

> *Note: Trinket is governed by the terms and conditions of the company, Trinket. This book uses references and screenshots with permission from Trinket.*

Now that you have your Python editor set up, let's first explore the different ways you can use these tools to run your code.

Running a Python Program

Now that you have installed Mu Editor, we should look at different ways you can run Python code. But before we do that, let us understand what a script is, as we will be using it quite often in the rest of the book.

What Is a Script?

A script is a short and simple set of instructions saved in a text file that tells a computer what to do step by step. Even though they are often used interchangeably, we do not want to confuse "script" with "code." Code is a broader term. It can be a small set of instructions like a script or as large as an entire video game. Scripts are small and focus on one task at a time, like sending a text message when pizza is ready.

The act of writing a script is called *scripting*. A *scripting language* is a programming language used for scripting. Scripting languages are usually run by an interpreter, which reads and executes each instruction on the spot. Python is one such language, so its programs are often called scripts. However, Python programs can also be much more complex than simple scripts.

The script in the Mu editor screenshots is a simple script that prints "Hello, World!" when executed. We will see in the next section how to run this script.

Python interpreter can run code in two modes:

Interactive or REPL Mode

This is when you talk to Python interactively. REPL is pronounced as "repple," rhyming with "apple." It's an acronym for Read, Evaluate, Print, Loop. These are the four things that computer does in interactive mode:

1. Read the user input (this is your Python command).
2. Evaluate the code (Python interpreter run / executes your code).
3. Print results (this is the computer's response to show you the results of Step 2).

4. Loop back to step 1 (to continue the conversation).

In Mu Editor, you can start the REPL (or interactive mode) by clicking on the "REPL" in the top menu. When you are in an interactive mode, Python starts by giving a welcome message with its version number and copyright notice. Then it tells you that it's waiting for your instructions (code) by showing you either >>> or a numbered prompt `In [1]:`. The number one in the square bracket is the current line number. If you enter a code and press enter, the computer will show the output in line `Out[1]:` and move on to line number [2], and you will see `In[2]:`.

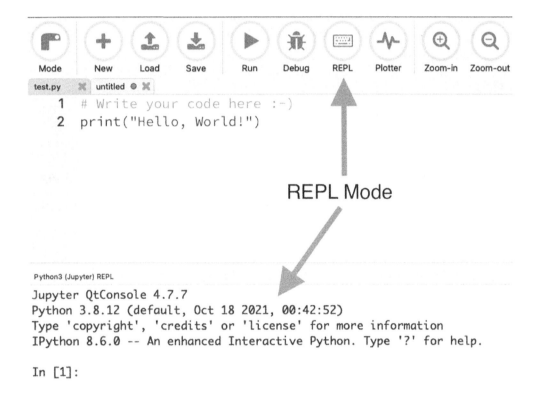

But why use this mode? Here, you can type the code directly and see the execution results instantly. This mode is an excellent way to learn, test code snippets, and get quick feedback without requiring a complete script or program.

Think of it like a notebook where you can quickly jot down ideas and try different things. Since the REPL gives instant feedback, it's quick and easy to experiment, explore, and see how the computer responds to your code.

Let us use this mode to write a one-line script that adds two numbers 2 and 3 and should print the result 5 when executed (by pressing the enter key).

```
Python3 (Jupyter) REPL
Jupyter QtConsole 4.7.7
Python 3.8.12 (default, Oct 18 2021, 00:42:52)
Type 'copyright', 'credits' or 'license' for more information
IPython 8.6.0 -- An enhanced Interactive Python. Type '?' for help.

In [1]: 2 + 3
Out[1]: 5

In [2]:
```

As shown in the above screenshot, the script was executed, the result was printed, and the Python interpreter is now ready for your next command, as indicated by `In [2]:`.

To exit the REPL (interactive mode), click the REPL button in the top menu again.

Script Mode

This is when you run the Python code saved in a file with a **.py** extension. The `.py` extension tell the Python interpreter that it's a Python file. This mode is good for larger programs.

You already used the script mode earlier in Fig. 1.2 when you saved a script named test.py and clicked the Run button on the top menu to execute it.

> Note: In this book, we will use both interactive and script modes wherever necessary. All the code in this book uses Python version 3.8.12. You should be able to run these codes if you have Python version 3.8 or higher installed on your system.

Activity: Explore Order of Operations Using REPL Mode. Let us do some hands-on exercises to explore Python's REPL (or interactive) mode and have some fun.

Open an REPL shell as discussed in the previous section. Now, try running the following mathematical expressions to observe the order of operations (PEMDAS—Parentheses, Exponents, Multiplication/Division, Addition/Subtraction) in action using Python.

Activity	Expression	Expected Result
Parentheses First	(5 + 3) * 2	16
Without Parentheses	5 + 3 * 2	11
Exponents	2 ** 3 + 1	9
Multiplication and Division Together	12 / 4 * 3	9.0
Addition and Subtraction Together	10 - 2 + 4	12

Code Formatting in the Book

Throughout this book, we will use code blocks to present the example code and their corresponding output. Please note that the output appears below the dotted line labeled `----Output-----------------` as shown below.

```
x = 2
x += 5        # Adds and Assigns
print("x = ", x)
----Output----------------
x = 7
```

This format helps you easily distinguish between the **example code** and the **output** generated by that code.

> **Important Note:**
>
> Please note that **do not copy the code below the output line for execution**. The output is intended as a reference to help you understand what to expect when running the code.

Review Questions

Read the questions carefully and select the correct options to assess how much you have learned so far. Note that more than one answer may be correct.

1. Which characteristic of Python makes it a preferred choice for beginners?
    a. Its complex syntax
    b. Its slow execution speed
    c. Its simple, English-like syntax
    d. Its requirement for a compilation step

2. What is an advantage of interpreted languages like Python?
    a. Faster development
    b. Requires separate compilation
    c. Slower development
    d. Hard to debug

3. What is a script in the context of computer programming?
    a. A short and simple set of instructions
    b. Code for mobile app
    c. Complex set of rules
    d. A long running program

4. What does Python's interactive mode (REPL) allow you to do?
    a. Write a code in a file
    b. Run code directly and see results instantly
    c. Compile code

d. Build an app
5. What is a compiled language characteristic that interpreted languages like Python do not share?
   a. Instant execution of code line by line
   b. Easy to debug
   c. Immediate feedback
   d. Need for a separate compilation step

6. Which mode is used to run a complete Python script saved in a file?
   a. Interactive / REPL mode
   b. Debugging mode
   c. Script mode
   d. Compilation mode

7. What does the Python script "print('Hello, World!')" do?
   a. Instructs computer to calculate
   b. Writes "Hello, World!" in the code
   c. Prints "Hello, World!" to the screen
   d. Sends an email with a "Hello, World!" message

Answer Keys:
1. c; 2. a; 3. a; 4. b; 5. d; 6. c; 7. c

2

Drawing with Python

Introduction to Turtle Graphics

Before we get into the detailed concepts of Python programming, let's first start thinking like a programmer. What does it mean to think like a programmer? It means breaking down problems into smaller parts and telling the computer exactly what to do step by step.

In this chapter, we'll use Turtle graphics, a fun and engaging way to learn Python. It lets you create colorful graphics and animations by controlling a turtle that moves around the screen based on simple Python commands. You'll need to give the turtle clear instructions to draw shapes, lines, and designs. While it's not a real turtle, it's a drawing tool that moves around the screen and follows your commands.

> Note: If you're not a fan of visual programming—perhaps because you've done a lot of Scratch programming in school—and you're more interested in text-based Python, feel free to skip this chapter and move directly to Chapter 3. However, if you'd like to explore the Turtle approach to learning Python, keep reading.

We'll write code to control the turtle's movement and create cool designs with just a few simple commands.

Let's get started. Open your Mu editor, and let's write our first piece of code:

```
1. import turtle
2. t = turtle.Turtle()
3. t.forward(100)
```

Run this code first, and then we'll break it down to understand what it does.

Line #1 loads the Turtle module into the computer's memory. We'll learn about modules, classes, methods, etc., in later chapters. For now, just know that this library provides everything you need to create and control the turtle. The Turtle module is a built-in Python library that allows you to create graphics and draw shapes on the screen.

Line #2 creates the turtle (or pen) that we'll use to draw. We've named this pen t, but you can name it anything you like—your name, your favorite pet's name, or even a random word.

Line #3 tells the turtle to move forward by 100 units and draw a line while moving.

The output screen should look like below:

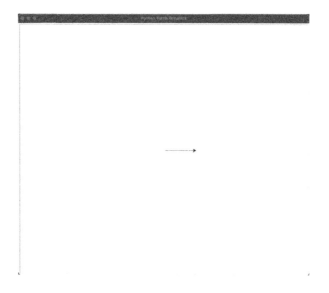

Now that you've moved the turtle forward, let's see how we can make it turn. But wait, you might be wondering: what do I mean by "turn"? How much should it turn? When you ask someone to turn, they might make a full turn, a slight turn, or a right-angle turn. In computer programming, we need to be very specific.

Building on the previous code, let's explicitly tell the turtle to turn 90 degrees. But even then, "turn 90 degrees" isn't specific enough. Can you see why? The turtle has two possible options: it could turn 90 degrees to the left or 90 degrees to the right, and it doesn't know which one to choose. So, let's be precise and tell the turtle to turn 90 degrees to the left, then move forward 100 units.

```
t.left(90)
t.forward(100)
```

Degrees

If you have not learned degrees yet, let me explain it briefly. Those who already know can skip to the next section.

If you stand at the center of a circle, and there's a line pointing straight in front of you (like 12 o'clock on a clock). You can say that line is pointing "0 degrees."

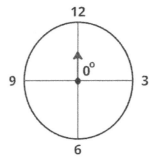

If you turn to your right, you move in a circle. As you turn, you are moving through degrees.

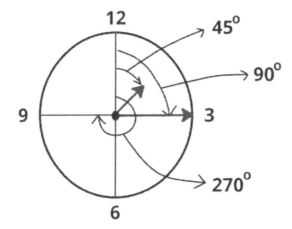

A **degree** is simply a way to measure how far you've turned from your starting position. We divide a full circle into 360 equal parts, and each part is called a degree. So, if you turn all the way around and end up in the same spot, you've turned 360 degrees. If you turn just a little, say 90 degrees, you're now facing 3 o'clock, because 90 degrees is one-quarter of the full circle. On a clock, the hour hand moves 30 degrees every hour.

One thing you should observe – whether you instruct your turtle to turn left 90 degrees or turn right 270 degrees, it will end up in the same position. Can you tell why?

There are many useful turtle commands you should learn to experiment with, so you can see how your instructions are translated into output on the screen.

- `forward(distance)`: Move the turtle forward by the specified distance.
- `backward(distance)`: Move the turtle backward by the specified distance.
- `right(angle)`: Turn the turtle right by the specified angle.
- `left(angle)`: Turn the turtle left by the specified angle.

Now, let's see how we can change the pen we're using to draw. By default, the pen color is black, but what if you want to draw with a different color, change the thickness of the pen, or fill a shape with a color? We have the following commands in Python to help us do all of these.

- `pendown()`: Lower the pen to start drawing.
- `penup()`: Lift the pen to stop drawing.
- `pensize(width)`: Set the width of the pen.
- `pencolor(color)`: Set the pen color.
- `fillcolor(color)`: Set the fill color.
- `begin_fill()`: Start filling the shape.
- `end_fill()`: Stop filling the shape.

Run the below code and see what it does:

```python
import turtle

# Create a turtle object
t = turtle.Turtle()

# Set the pen size to 3
t.pensize(3)

# Set the pen color to blue
t.pencolor("blue")

# Set the fill color to yellow
t.fillcolor("yellow")

# Begin filling the shape with the fill color
t.begin_fill()

# Lower the pen to start drawing
t.pendown()

# Draw a square
for _ in range(4):
    t.forward(100)
    t.right(90)

# Complete filling the shape
t.end_fill()

# Lift the pen to stop drawing
t.penup()

# Move the turtle to a new location
t.goto(-150, 0)

# Set the pen color to red
t.pencolor("red")
```

```python
# Set the fill color to green
t.fillcolor("green")

# Begin filling the shape with the fill color
t.begin_fill()

# Lower the pen to start drawing again
t.pendown()

# Draw a triangle
for _ in range(3):
    t.forward(100)
    t.left(120)

# Complete filling the shape
t.end_fill()
```

The output should look like the image below, with the square filled in yellow and the triangle in green.

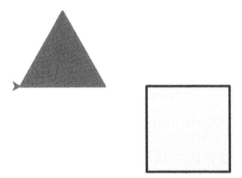

Try the code above by adjusting the values of the parameters in each command to see how the output changes.

Did you notice how we used the code `for _ in range(4):`? This is a loop that repeats the code inside it. You'll learn more about loops in later chapters, but for now, just understand that `for _ in range(5):` will execute the indented code below it 5 times.

For example, to create the square, we used `for _ in range(4):` because a square has 4 sides. Similarly, to draw the triangle, we used `for _ in range(3):` since a triangle has 3 sides.

With Turtle graphics, you can draw complex shapes and create limitless possibilities. With a turtle pen in hand and a blank canvas, you can transform your imagination into fascinating designs.

Take a look at the spiral below.

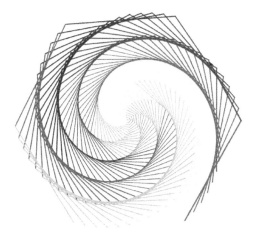

Does it seem complex to you? But did you know that the code used to create this is actually very simple?

```
1.  import turtle
2.  colors = ['red', 'purple', 'blue', 'green', 'orange', 'yellow']
3.  t = turtle.Turtle()
4.  turtle.bgcolor('black')
5.  for x in range(360):
```

```
6.      t.pencolor(colors[x%6])
7.      t.width(x//100 + 1)
8.      t.forward(x)
9.      t.left(59)
```

Let's understand how this code works:

Line #2 is a list of colors. We'll use these colors to create the rainbow effect in our spiral.

Line #5 defines a loop that runs repeatedly for 360 times. All the code below this line is part of the loop because it's indented.

Line #6 changes the color of the turtle's pen (the color used to draw). The expression `x % 6` calculates the remainder when x is divided by 6. This ensures that as x goes from 0 to 359, the color repeats in a cycle. When x is 0, `x % 6` is 0 the color is 'red' from the list of colors. When x is 1, `x % 6` is 1 the color is 'purple' and so on, cycling through the colors as `x` increases.

Line #7 controls the thickness of the pen. The expression `x // 100` divides x by 100 and gives the integer division result (which you will learn about in later chapter). This ensures that for x between 0 and 99, `x // 100` equals `0`, so the pen width will be 1. For x between 100 and 199, `x // 100` equals 1, so the pen width will be 2 and so on.

If this concept seems a bit tricky, don't worry. You can skip it for now and come back to it after you've learned more about these topics in later chapters.

Activity: Let's use the Turtle graphics module to draw a hexagon (a six-sided shape). The goal is to understand how the turtle moves and turns, and how to create geometric shapes by combining basic commands.

First, try it yourself, and then compare your code with the following solution.

```python
import turtle
t = turtle.Turtle()

num_sides = 6
side_length = 70
angle = 360 / num_sides

for i in range(num_sides):
    t.forward(side_length)
    t.right(angle)
```

Notice that the code `angle = 360.0 / num_sides` calculates the angle between each side of the hexagon. Since a hexagon has 6 sides, the angle is `360 / 6 = 60` degrees.

3

Variables

Labeling and Storing Data

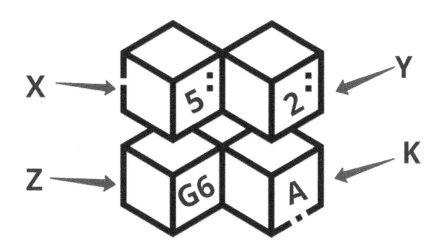

When you run your code, the Python interpreter reads it line by line. But what happens if it's currently reading line #5 and needs some information from line #1? To handle situations like this, Python stores data from your code in the computer's memory. Whenever Python needs the data, it retrieves it from memory. Once it's done with the data, it discards it to free up space for other things. The computer's memory can store billions of bits of data, and your program needs to quickly find and use this data to perform calculations. But how does it remember where to find this data? Let's explore this with an example.

Imagine you have three friends: Henry, Amelia, and Miles. You're planning a trip with them and decide to bring some candies to share. You want to give 3 candies to Amelia, 2 to Henry, and 4 to Miles. All the candies are the same, and you're worried you might forget how many to give to each friend. What could you do?

Here's one idea: You could use three boxes, label them with your friends' names, and put the right number of candies in each box. That way, when you meet them, you don't have to remember who gets what—you just open the box with their name on it.

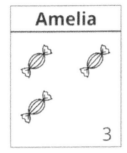 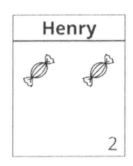 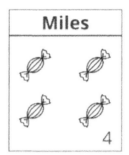

Now, let's make things more interesting. Amelia suddenly says, "I want as many candies as Miles and Henry combined!" What do you do? Here's a simple plan:

1. Open Miles's box and count the candies in it.
2. Open Henry's box and count the candies there.
3. Add the two numbers together.
4. Take out the candies from Amelia's box and replace them with the new total from Step 3.

Now, Amelia's box has 6 candies instead of the 3 she had before.

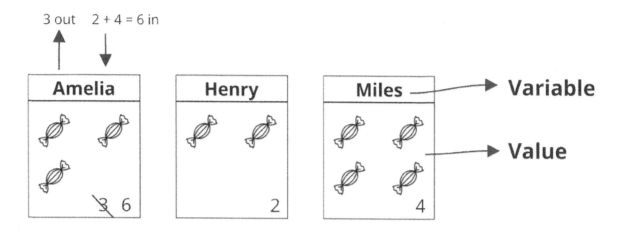

This is exactly how Python processes information. The boxes are spaces created in memory, the labels on the boxes are called **variables**, and the items you put inside the boxes are called **values**. Now, remember the question we asked earlier: "But how does it remember where to find this data?" Python does it just like you did by remembering the labels on the boxes. It uses the variable names to find and work with the data stored in memory.

If you had to write Python code for the example above, it would look like below. Try writing it in the Mu Editor and see what it prints.

```
# Initially you assign the values as below
amelia = 3
henry = 2
miles = 4

# Now, you modified as requested by Amelia
amelia = henry + miles

print(amelia, henry, miles)
----Output-----------------
```

To sum it up, a variable in Python is like a labeled box in your computer's memory where you can store a value. You can use this variable as many times and in as many places in your code as you want, and Python will know where to find it. For example, if you write `x = 2` and then, on another line, you write `y = x + 5`, the Python interpreter will look up the value of x (which is 2), add it to 5, and store the result in a new variable called y. Isn't that awesome? Python makes it super easy to work with variables and do calculations.

To summarize:

- Variable is a location in the computer's memory where you can store a value and reference it later.
- You can update the value stored in a variable any time you need to. Assigning a new value to an existing variable replaces the previously stored value.

Variable Assignment Statement

A variable assignment statement creates a new variable and assigns it a value.

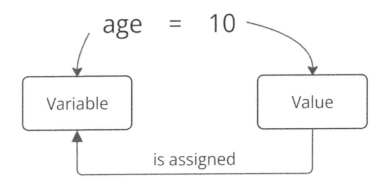

This statement tells the interpreter to store the value on the right-hand side in the memory location associated with the variable name on the left-hand side. For instance, in the statement `age = 10`, the integer 10 is assigned to the variable age.

Variable Naming Rules

Can we name variables whatever we want? The answer is no. Python has some basic rules that must be followed when naming a variable. Python doesn't care what name you assign to a variable if it adheres to these rules:

1. Variable names must start with a letter (A–Z, uppercase or lowercase) or an underscore (_)
   This means a variable name cannot start with a number or any other symbol.
2. Variable names can only contain letters, numbers, and underscores.
   This means that after the first character (which must be a letter or underscore by Rule# 1), the remaining characters in the variable name can be any combination of letters, numbers, and underscores.
3. Variable names are case sensitive.
   This means that name and Name are two different variables.
4. Python has some reserved words, or keywords of the language (Table 3.1), and those keywords cannot be used for naming a variable.

Examples of valid variable names: `age, Name, _address, phone_1`

Examples of invalid variable names: `2address, $test, @age, -phone, class, TRUE`

FALSE	await	else	import	pass
None	break	except	in	raise
TRUE	class	finally	is	return
and	continue	for	lambda	try
as	def	from	nonlocal	while
assert	del	global	not	with
asynch	elif	if	or	yield

Table 3.1 Python Reserved Keywords

In addition to the strict rules for naming variables, there are some helpful conventions that most Python developers follow. These conventions aren't rules that Python forces you to follow, but they're widely used because they make your code easier to read and understand. Following these conventions also helps others understand your code more quickly.

1. In Python, it's a good idea to write variable names using snake case (https://en.wikipedia.org/wiki/Snake_case). This means all the letters are lowercase, and words are separated by underscores, like this: `user_address` or `number_of_children`.
2. When naming variables, try to keep the name short yet descriptive. For example, using a name like score makes it clear what the variable represents, while a name like s, though shorter, can be confusing because it doesn't provide much context about what it stands for.

You can assign multiple variables in one line.
For example, `a, b, c = 1, 2, 3` assigns 1 to a, 2 to b, and 3 to c.

Review Questions

Read the questions carefully and select the correct options to assess how much you have learned so far. Note that more than one answer may be correct.

1. Which of the following variable names is NOT allowed in Python?
    a. my_name
    b. score$
    c. userName
    d. 5th_place

2. What will be the value of variable y in the following code?

    ```
    x = 2
    y = x
    x = 7
    ```

    a. 2
    b. 7
    c. Error: x cannot be reassigned
    d. None

3. Which of these is a Python keyword?
    a. while
    b. output
    c. calculate
    d. result

4. What does the following code do?

    ```
    score = 50
    score = score + 10
    ```

a. Creates a new variable with the value 10
b. Adds 10 to the variable score and stores the result back in score
c. Error: Invalid operation
d. Both a and b

5. Which of the following variable assignments is invalid in Python?
    a. age = 21
    b. height = 8
    c. is_student = True
    d. continue = 1

6. Which of the following are true about Python variables?
    a. Variables in Python must be declared with a specific type.
    b. Variable names can include letters, numbers, and underscores, but cannot start with a number.
    c. Python variables are case-sensitive.
    d. You must assign a value to a variable before using it in Python.

7. Which of the following are valid ways to declare a variable in Python?
    a. x = 10
    b. 10 = x
    c. x = "Hello"
    d. int x = 10

Answer Keys:
1. b and d; 2. a; 3. a; 4. b; 5. d; 6. b, c, and d; 7. a and c

4

Function Basics

The Basics of a Magical, Relentless Helper

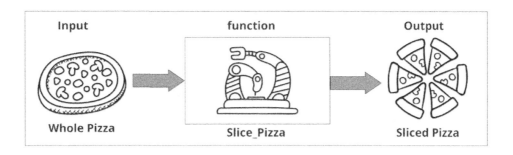

A **function** is a set of reusable code designed to perform a specific task repeatedly. Let's understand this concept further with an analogy.

At the pizza store, you must have seen someone using a slicer to cut the pizza into pieces. Have you noticed how uneven those slices can be? What if you need to create a machine (as shown in the above diagram) that does this job? You could feed a pizza of any size into this machine and instantly get a perfectly sliced pizza. Let's call this machine `slice_pizza`. This is exactly how Python functions work. They perform repetitive tasks by encapsulating reusable code that gives an output for a given input. Every time you use these functions, you don't need to worry about what's inside. All that matters is the input and the output. In this example, the whole, unsliced pizza is the input, and the evenly sliced pizza is the output. Using the function to produce the output is called "invoking" or "calling" the function. Thinking from a coding perspective, if `slice_pizza` is a Python function, then you call it by suffixing it with an open and close parenthesis like `slice_pizza()`.

A function may or may not expect inputs. If it does not, you simply open and close the parentheses when calling the function. However, if the function expects inputs, those are passed within the parentheses. These inputs, known as **parameters**, can be one or more. When you call a function with parameters, the values you provide are called **arguments**. This process of providing arguments is how you "pass" them to the function. So, simply speaking, a function takes parameters, and we pass arguments when calling the function. In the function code, it maps the supplied arguments to its parameters. People often use "parameter" and "argument" interchangeably, but I want you to learn the difference.

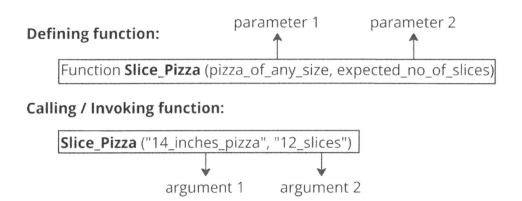

Python's Built-in Functions

Python has many pre-made functions to perform common tasks, and they are ready for you to use. These are called Python's **built-in functions**. They are pre-packaged to save time and effort because you don't have to write the code yourself. Python has already done it for you. One of the most common built-in functions is `print()`, and we will learn all about it next.

Printing Python's Output – The `print()` Function

This function allows you to display various types of information on the screen. Figure below shows the `print()` function taking the text `Hello, Python!` as input and doing all the work behind the scenes to display it on the screen.

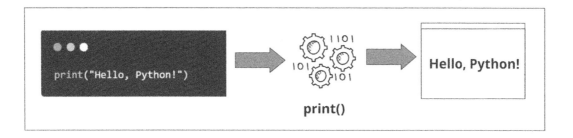

The `print()` function can help you retrieve the value of a variable from computers memory and show it on your computer screen. Look at the following code and see how it's performing this task.

```
first_name = "Amelia"
last_name = "Smith"
print(first_name, last_name)
----Output----------------
Amelia Smith
```

We learned that in Python, the `print()` function is used to display messages or show information on the screen. In most of the examples above, we have used this function. For example, if you write `first_name = "Amelia"`, Python stores the value `"Amelia"` in a variable `first_name` of type string. However, if you want to retrieve this value from memory and display it to the user, you need to use the `print()` function.

The `print()` function can be used in many ways. Below, I will list some of them.

Printing multiple things at once

The `print()` function can print multiple items on the same line. You just need to separate them with commas, as shown below. When you print multiple items in a single `print()` statement, Python automatically separates them with a space.

```
first_name = "Amelia"
age = 13
print(first_name, "is", age, "years old. ")
----Output----------------
Amelia is 13 years old.
```

sep Parameter

The `print()` function has an optional parameter called **sep**. As we saw in the code above, when we print multiple items at once, separated by commas, Python, by default, inserts a space between them. However, we can change this behavior by providing an argument for the `sep` parameter. See the code below. In this example, instead of the default space, we passed the argument `sep="*"`, and Python used the asterisk as the separator between the items.

```
first_name = "Amelia"
```

```
age = 13
print(first_name, "is", age, "years old. ", sep="*")
----Output----------------
Amelia*is*13*years old.
```

How can you print something with no spaces in between? Let's look at the example in the code below. In this example, we passed an empty string as the `sep` argument.

```
first_name = "Amelia"
age = 13
print(first_name, "is", age, "years old. ", sep="")
----Output----------------
Ameliais13years old.
```

end Parameter

The `print()` function takes another optional parameter called end. This parameter decides what happens after the `print()` finishes. By default, Python adds a new line at the end of each print statement. This moves the cursor on the next line so that the next thing you print gets printed on the next line. But you can change this behavior by sending an argument for the end parameter. You can make it print something else, or even nothing at all. Try running the code shown in below using your Mu Editor.

```
first_name = "Amelia"
last_name = "Smith"
print(first_name, last_name, end=" ")
----Output----------------
Amelia Smith >>>
```

Here, we used " " (a space) as the end parameter. The first `print()` ends with a space instead of a new line, so the second `print()` continues on the same line. You can also see the prompt >>> is on the same line this time.

Repeating Print with String Multiplication

You can repeat a string multiple times by using the multiplication operator *. This is useful for generating patterns or printing something multiple times. See the code below. Try experimenting with it in Mu Editor.

```
first_name = "Amelia"
last_name = "Smith"
print(first_name * 2, last_name * 3)
print("^" * 32)
----Output----------------
AmeliaAmelia SmithSmithSmith
^^^^^^^^^^^^^^^^^^^^^^^^^^^^^^^^
```

Receive User Feedback – The `input()` Function

Would you ever need to ask the user for any information to run your code? What if you want to write a program that asks the user for their name and then prints a customized greeting with their name in it?

Python has a built-in function called **input()**. The `input()` function in Python allows you to ask the user for information. It's like asking a question, and the user types an answer that the program can use later. When your program runs, it will stop and wait for the user to type the requested information on the keyboard. After the user types their answer and presses Enter, the program resumes.

Look at the example in code below. Can you try this in your Mu Editor and see how it works?

```
user_input = input("what is your name?")
print("Hello, " + user_input + "!")
----Output----------------
what is your name?Anand
Hello, Anand!
```

Here is how it works:
1. The `input()` function will display the message inside the quotation marks (in this case, `What is your name?`).
2. The program will pause and wait for you to type something.
3. Once you type your answer and press Enter, the program saves whatever you typed as `user_input`.
4. The program prints `Hello,` followed by the data you entered. In this case, I entered my name.

Review Questions

Read the questions carefully and select the correct options to assess how much you have learned so far. Note that more than one answer may be correct.

1. What is the main purpose of a function in Python?
    a. To store data in memory
    b. To perform a specific task repeatedly
    c. To create new variables
    d. To pass arguments to map to parameters

2. In the analogy of the pizza store, what does the "slicer" represent in Python?
    a. The variable
    b. The function
    c. The argument
    d. The output

3. In the context of functions, what is the difference between parameters and arguments?
    a. Parameters are values passed to the function, while arguments are the names used in the function
    b. Parameters are only used with built-in functions, and arguments are used with user-defined functions
    c. Arguments are values passed to the function, while parameters are the names used in the function
    d. There is no difference

4. What happens when you call the `input("Your name?")` function in Python?
    a. It displays a message on the screen
    b. It waits for the user to type an answer
    c. It saves the typed answer to a variable
    d. All of the above

5. What is the primary function of the `print()` function in Python?
    a. To create new variables
    b. To take user input
    c. To display information on the screen
    d. To perform calculations

6. Which of the following statements are true about the function `input()` in Python?
    a. input() always returns the user input as a string.
    b. The input() function automatically converts the input to an integer if you enter a number.
    c. The input() function pauses the program execution until the user provides input.
    d. You cannot pass a message inside the parentheses of `input()`.

7. What will be the output of the following code?

```python
first_name = "John"
age = 25
print(first_name, age, end="")
```

a. John 25
b. John25
c. John 25 followed by a newline.
d. John25 followed by a newline.

8. What happens when you pass an empty string `sep=""` as a parameter in the `print()` function?
    a. It removes all spaces between the printed items.
    b. It adds extra spaces between the printed items.
    c. It prints the items without any space between them.
    d. It will raise an error.

9. What will be the output of the following code?

    ```
    first_name = "Amelia"
    last_name = "Smith"
    print(first_name * 2, last_name * 3)
    ```

    a. AmeliaAmelia SmithSmithSmith
    b. AmeliaAmelia Smith Smith Smith
    c. Amelia Amelia Smith Smith Smith
    d. AmeliaAmeliaSmithSmithSmith

Answer Keys:
1. b; 2. b; 3. c; 4. d; 5. c; 6. a, c; 7. a; 8. c; 9. a

5

Data Types

Categorizing the Data

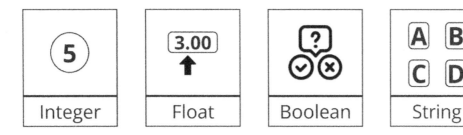

Earlier, we saw (in the chapter on variables) that Amelia labeled boxes with her friends' names. Now, imagine she needs to give her friends not just candies, but also things like pencils, marbles, and more. She wouldn't put marbles and pencils in a box meant for candies, right? That would make it hard for her to find things. One solution is to use boxes of different colors, with each color only holding a specific type of item. For example, all the candies go in red boxes, pencils go in blue boxes, and marbles go in green boxes. This way, Amelia can organize everything more easily and find what she needs faster because she not only knows what's written on the label but can also tell the type of box by its color.

This is like what we call a "data type" in Python. Just like each colored box can only hold a certain type of item, in Python, a data type can only hold a certain type of value. For example:

- Integer box: This box holds whole numbers like your age (13), the number of goals you scored (2), or the number of points you earned in a game (100).
- Text box: This box holds words and letters like your name ("Amelia"), your favorite song, or a secret message.
- Decimal box: This box holds numbers with decimals, like the price of a toy ($7.99), your height (1.65 meters), or your score in a race (12.34 seconds).

These data types help Python understand and process information correctly, just like Amelia needs the right kind of box to keep her things organized and easy to find. We can call these data types, a category for a value. Let's explore some of these data types.

Integers

Integers are whole numbers (negative, positive, or zero) without a decimal point. Python calls this data type "**int**."

Examples: `-100, 4, 1564, 43543, -7622, 0`

Is `-100,000` an integer data type? The answer is no. In Python, numbers are considered integers (int) if they are written **without any commas** or other non-numeric characters. So, the correct representation of negative one hundred thousand as an integer would be -100000.

Python has a special built-in function called **type()**. This function returns the data type of a variable or value. You can pass a variable or value as an argument to `type()`, and it will return the type of the object. Let's check it out with the following example code.

```
my_age = 12
type_of_my_age = type(my_age)
print(type_of_my_age)
----Output----------------
<class 'int'>
```

In this example:
1. We have a variable `my_age`, and we assign the value 12 to it on line #1.
2. On line #2, we call the `type()` function, passing `my_age` as an argument. The function returns the type of `my_age`, which is assigned to another variable named `type_of_my_age`.
3. Finally, we print the value of `type_of_my_age` to see the type.

When you run this code, Python returns `<class 'int'>`, indicating that the type of `my_age` variable is int (integer).

Floating Point Numbers

Floating Point Numbers are numbers that contain a decimal point. Python calls this data type "float". The term "float" comes from "floating point number," where the decimal point can "float" to any position within the number.

Examples: `-100.00, 4.56, 156.4, -0.043543, -7.622, 0.0, -1.0`

Note: The values 132.000, 0.0, and -1.0 are all considered floats. In Python, any number that includes a decimal point is treated as a float, even if the decimal part is zero.

Is -0.0 a valid float? Why or why not? Yes, it's a valid float (see the code below). In Python, -0.0 is treated as a valid floating-point number, distinct from 0.0, although they are numerically equivalent when it comes to calculations. -0.0 and 0.0 are both floating-point representations of zero.

```
my_score = -0.0
print(type(my_score))
----Output----------------
<class 'float'>
```

Strings

Strings are sequences of characters. You can think of a string as a collection of characters, like beads on a necklace. Characters can be letters, numbers, symbols, or spaces. Strings are enclosed in either single quotes (' ') or double quotes (" "). Python calls this data type "str".

Examples: `"Hello, World!"`, `'12345'`, `'Learning is fun 4 sure!'`, `"30% attendance"`, `"-3.0"`, `'200'`, `"6 - 2"`

In Python, any text surrounded by quotation marks ("") or single quotes (' ') is treated as a string, even if the text inside consists of integers, decimals, special characters, or anything else. As long as the text is enclosed by quotes or quotation marks, it is considered a string. Let's look at some valid and invalid strings to make it clearer. Note that you cannot mix single quotes and double quotes within the same string, as shown in the examples below.

Valid string	Invalid string
"Python is fun!"	Python is fun!
'I like Python'	'I like Python"
"Let's go to school"	'Let's go to school'
"100"	100
'He says, "I know Python."'	"He says, "I know Python.""

Are "" and " " strings? Yes. Python considers both of them as strings (see the code below). "" is called an empty string, and " " is a string that contains one character: a single space.

```
string_1 = ""
string_2 = " "
print(type(string_1))
print(type(string_2))
----Output----------------
<class 'str'>
<class 'str'>
```

Strings are an interesting data type because they support a variety of operations and offer many built-in functions. In a later chapter, we will explore strings in more detail, covering topics such as creating strings, comparing string values, string concatenation, slicing, searching, testing, formatting, and modification. But for now, let's go over some of the basics to get started.

Boolean

Boolean is a type of data that can only be one of two possible values:

- True
- False

Booleans are often used in conditions to make decisions (like checking if something is true / false, on / off or yes / no). Python calls this data type "bool".

The capitalization is very important here. Booleans must be capitalized as **T**rue and **F**alse. Python will throw an error if you use lowercase **t**rue or **f**alse for a Boolean variable.

Do you know why it's called Boolean? The name comes from a mathematician named George Boole, who created a system of logic called Boolean logic. This system helps computers make decisions and solve problems by asking questions with only two answers: true or false.

Example: Look at the code below. Here, say for a restaurant, `is_open` is set to *True*, meaning it's open. `is_serving` is set to *False*, meaning it's not serving food at this moment.

```
is_open = True
is_serving = False
print(is_open)
print(is_serving)
----Output----------------
True
False
```

Booleans are very important because they are used everywhere in programming. In later chapters, when we learn about conditionals and loops, we will discuss Booleans in more detail.

Other Data Types

There are other important data types, such as Sequences (Lists and Tuples), Mappings (Dictionaries), and more. These are more advanced concepts, and I plan to cover them later, once you've built a solid foundation with the basics.

Data Type Conversion

Consider a scenario where the user inputs a number as a string, say "2". What if you want to add this user input to another number, say 5? Can you do that? Would `"2" + 5` work? The answer is no. It will result in an error, as shown in the below code. In Python, it's possible to convert a value from one type to another. In this example, we can easily convert the user-entered string to a number. Let's look at some common type conversions.

```
score = "50"
new_score = score + 10
print(new_score)
----Output----------------
Traceback (most recent call last):
  File "/Users/anandpandey/mu_code/test1.py", line 2, in <module>
    new_score = score + 10
TypeError: can only concatenate str (not "int") to str
```

Converting Strings to Integers

Python's built-in function **int()** converts a number stored as a string back into an integer. The code below shows the corrected version of the previous section's code using `int()`.

```
score = "50"
new_score = int(score) + 10
print(new_score)
----Output-----------------
60
```

Can you explain why the below code throws an error?

```
score = "50a"
new_score = int(score) + 10
print(new_score)
----Output-----------------
Traceback (most recent call last):
  File "/Users/anandpandey/mu_code/test1.py", line 2, in <module>
    print(int(score))
ValueError: invalid literal for int() with base 10: '50a'
```

The error occurs because the string "50a" contains a non-numeric character ('a'), making it an invalid literal for conversion to an integer.

Converting Strings to Floats

Python's built-in function **float()** converts a string that has a decimal point into a float. Note it prints float 50.5 not the string "50.5" as shown below.

```
score = "50.5"
```

```
new_score = float(score)
print(new_score)
----Output-----------------
50.5
```

Converting Numbers to Strings

Python's built-in function **str()** converts a number to a string.

```
score = 50
print(score + 10)
new_score = str(score)
print(new_score)
print(new_score + 10)
----Output-----------------
60
50
Traceback (most recent call last):
  File "/Users/anandpandey/mu_code/test1.py", line 4, in <module>
    print(new_score + 10)
TypeError: can only concatenate str (not "int") to str
```

In the code example shown above, the first two `print()` statements work, but the third one throws an error. Do you know why? The first `print()` statement outputs the sum of two integers, 50 + 10, which equals 60. The second print() statement displays the string value `new_score = "50"` without any error. However, the third print() statement tries to add the string value `new_score` (which is "50") to an integer, causing an error due to a type mismatch.

Review Questions

Read the questions carefully and select the correct options to assess how much you have learned so far. Note that more than one answer may be correct.

1. What will be the output of the code below?

    ```
    x = 5
    y = 2.5
    print(x + y)
    ```

    a. 12.5
    b. Error
    c. 7.5
    d. 5.2

2. What type of data does the code `result = str(50) + " is a number"` return?

    a. int
    b. float
    c. str
    d. bool

3. What will happen if you try to add an integer and a string in Python as shown in the code below?

    ```
    num = 10
    text = "Hello"
    print(num + text)
    ```

a. It will print "10Hello"
b. It will throw an error
c. It will print "Hello10"
d. It will print nothing

4. Which of the following is a valid way to convert a string to an integer in Python?
    a. int("123")
    b. float("123")
    c. str(123)
    d. convert("123")

5. What type of data does the `input()` function return in Python?
    a. int
    b. bool
    c. float
    d. str

6. What will the following code output?

```
score = 50
print(type(score))
```

a. <class 'float'>
b. <class 'int'>
c. <class 'str'>

d. TypeError

7. Which of the following values will be accepted by the `int()` function without throwing an error?
    a. "50.5"
    b. "50a"
    c. "50"
    d. "3.14"

8. Which of the following statements is true when converting the string "3.14" to an integer using `int()` in Python?
    a. It will raise a `ValueError` because int() cannot handle decimals.
    b. It will round the number and return 3.
    c. It will throw a `TypeError` because you can't directly convert a string with a decimal point.
    d. It will return 3 without rounding, truncating the decimal part.

Answer Keys:
1. c; 2. c; 3. b; 4. a; 5. d; 6. B; 7.c; 8. a

6

Code Execution Flow

Visualizing How Python Executes Code

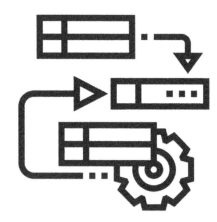

Do we need to understand how the Python interpreter executes a script? While it's not mandatory to know this, when we write a Python script, it's important to mentally execute each step line by line to ensure the script will produce the desired output when we run it through the Python interpreter. But how can we understand how Python actually executes a script behind the scenes? Let's take a closer look. We will take the example as shown below and run it step by step.

```
score = 10 * 2
new_score = score
print(new_score)
score = 40
new_score = score + new_score + 10
print("Total score is ", score + new_score)
```

Can you try running it line by line in your head? What do you get as the final output?

The Python interpreter takes the following steps to execute this code line by line:

STEP 1. Reads the first line.

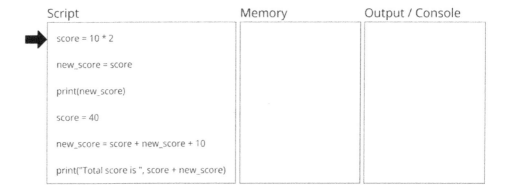

STEP 2. Simplifies the expression `10 * 2` to 20. (Note: We use the asterisk (*) for multiplication in programming.)

STEP 3. Since the variable score doesn't exist in memory, Python creates it and assigns the value 20.

STEP 4. Reads the next line.

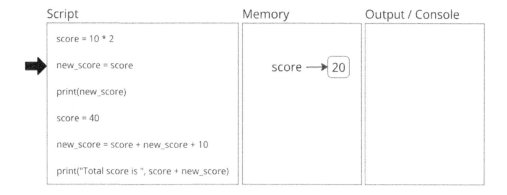

STEP 5. Retrieves the value of the variable score from memory: `score = 20`.

STEP 6. Since the variable `new_score` doesn't exist in memory, Python creates it and assigns the value 20.

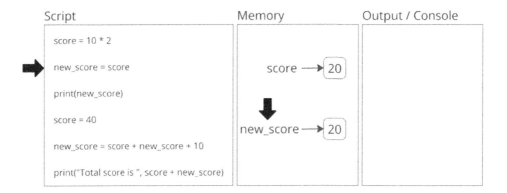

STEP 7. Reads the next line.

STEP 8. Retrieves the value of the variable `new_score` from memory: `new_score = 20`.

STEP 9. The print() function displays the output in the console.

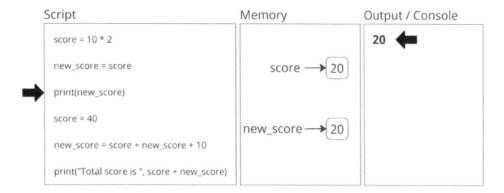

STEP 10. Reads the next line.

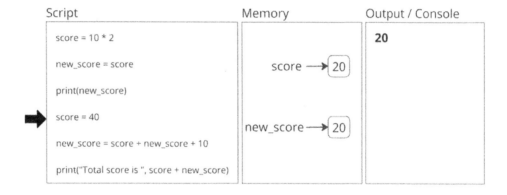

STEP 11. Since the variable score already exists in memory, Python updates its value by assigning the new value, 40.

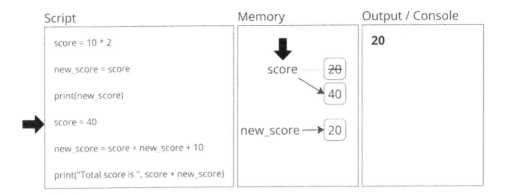

STEP 12. Reads the next line.

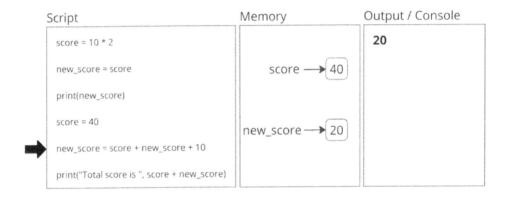

STEP 13. Retrieves the value of the variable score from memory: score = 40. Retrieves the value of the variable `new_score` from memory: `new_score = 20`.
Simplifies the expression `score + new_score + 10` to 40 + 20 + 10, which equals 70.

STEP 14. Since the variable `new_score` already exists in memory, Python updates its value by assigning the new value, 70.

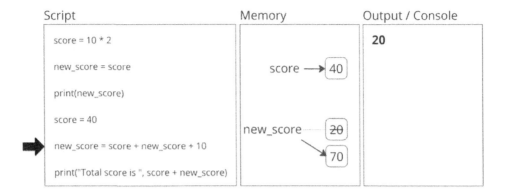

STEP 15. Reads the next line.

STEP 16. Retrieves the value of the variable score from memory: `score = 40`.
Retrieves the value of the variable `new_score` from memory: `new_score = 70`.
Simplifies the expression `score + new_score` to `40 + 70`, which equals 110.
Passes the simplified value 110 to the print() function for output.

STEP 17. The print() function displays the output in the console. After that, the script execution ends.

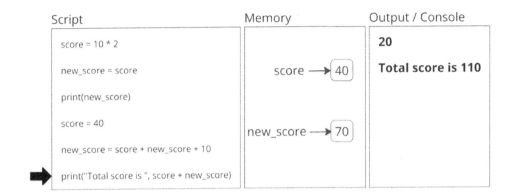

7

Operators

A Tool for Everyday Tasks

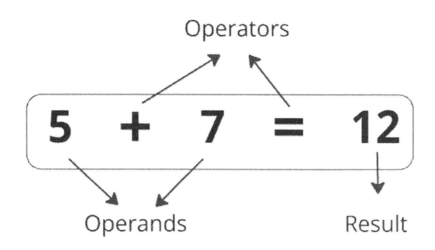

If you need to cut a piece of paper, what tool would you use? Scissors, right? Similarly, you use a knife for chopping vegetables, a spoon for eating, and so on. These are simple tools that help you with everyday tasks. In Python, we have similar tools called operators that help you perform small tasks in your code, like adding, subtracting, multiplying, or comparing numbers.

Operators are special symbols that carry out specific tasks on values or variables. An **operand** is the value or variable that the operator acts on. It's like the material or object you work with when using a tool. For example, in the expression 5 + 3, 5 and 3 are the operands, and + is the operator that tells Python to add them together.

We will briefly discuss different types of operators in this chapter to give you an idea of how many operators there are and what we need to deal with while coding. However, a detailed discussion of these operators is covered in later chapters when we discuss numbers, strings, lists, tuples, dictionaries, etc.

Types of Operators

There are many different types of operators in Python, each with its own meaning and purpose. We will explore some of the most common types that you will use frequently.

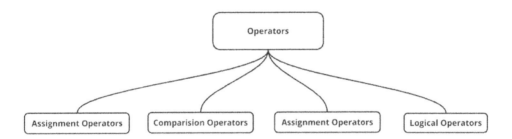

Arithmetic Operators

Python has mathematical operators that help us perform complex mathematical tasks.

Operator	Operation	Example	Result	Read as
+	Addition	5 + 3	8	5 plus 3
-	Subtraction	10 - 4	6	10 minus 4
*	Multiplication	7 * 2	14	7 times 2
/	Division	8 / 4	2	8 divided by 4
//	Floor Division	9 // 4	2	9 floor-divided by 4
%	Modulus (Remainder)	10 % 3	1	10 modulus 3 (remainder of 10 divided by 3)
**	Exponentiation	2 ** 3	8	2 raised to the power of 3

These are normal mathematical operators that you would have studied in your mathematics class. But some of them have special symbols which is special to Python like // for floor division and ** for exponentiation. There are some interesting facts here and we will discuss each of them in the chapter on numbers.

Comparison Operators

In Python, we often need to compare the values of two operands to determine things like which one is greater, which one is smaller, whether they are equal, or whether they are not equal. Comparison operators help us perform these comparisons.

Operator	Operation	Example	Result	Read as
==	Equal to	5 == 5	True	5 is equal to 5
!=	Not equal to	3 != 4	True	3 is not equal to 4
>	Greater than	7 > 3	True	7 is greater than 3
<	Less than	2 < 5	True	2 is less than 5
>=	Greater than or equal to	6 >= 6	True	6 is greater than or equal to 6
<=	Less than or equal to	4 <= 10	True	4 is less than or equal to 10

Let's say we have two integer variables, x and y, and we want to check if x is greater than y. We can use the ">" operator to perform this comparison, as shown in the code below. Try running this code in your Mu Editor by changing the values of x and y. What results do you get?

```
x = 17
y = 11
result_comparison = x > y
print(result_comparison)
----Output----------------
True
```

Also, try running the other operators and note the results in the code below.

```
x = 17
y = 11
print("x == y : ", x == y)   # Equal
print("x != y : ", x != y)   # Not Equal
print("x < y : ", x < y)     # Less Than
print("x > y : ", x > y)     # Greater Than
print("x <= y : ", x <= y)   # Less Than or Equal
print("x >= y : ", x >= y)   # Greater Than or Equal
----Output----------------
x == y :   False
x != y :   True
x < y :    False
x > y :    True
x <= y :   False
x >= y :   True
```

Note that == is used for comparison, not =. Why? Because = is used for assigning values to a variable. You'll see more about this in the next section.

Assignment Operators

Assignment operators are used to assign values to variables. You might be wondering, 'What else would you need, other than the = operator, for this?' Well, there are also special assignment operators that allow you to do more than just store a value in a variable as shown in the below table.

Operator	Operation	Example	Result	Read as
=	Assigns a value to a variable	x = 5	x = 5	x is assigned the value 5
+=	Adds and assigns	x += 3	x = x + 3	x plus equals 3
-=	Subtracts and assigns	x -= 2	x = x - 2	x minus equals 2
*=	Multiplies and assigns	x *= 4	x = x * 4	x times equals 4
/=	Divides and assigns	x /= 2	x = x / 2	x divided by equals 2
//=	Floor divides and assigns	x //= 3	x = x // 3	x floor divided by equals 3
%=	Modulo (remainder) and assigns	x %= 3	x = x % 3	x modulus equals 3 (remainder of x divided by 3)
**=	Exponentiation and assigns	x **= 2	x = x ** 2	x raised to the power of equals 2

These special assignment operators are particularly useful for quickly combining or modifying values. For example, suppose you have a variable x, and you want to assign the value 2 to it. You would write `x = 2`, right? But what if I asked you to add 5 to x and then update x with the resulting value? You would do something like `x = x + 5`.

However, the += operator allows you to avoid repeating x and simply write `x += 5`. This means exactly the same thing as `x = x + 5`, as shown in the code below.

```
x = 2
x += 5      # Adds and Assigns
print("x = ", x)
----Output----------------
x = 7
```

Similarly, the -= operator allows you to subtract a value from a variable and assign the result to the variable itself. If x was 2 before, `x -= 1` would result in x being updated to 1 (i.e., 2 - 1 = 1).

Chaining Assignments

If you have multiple variables and want to assign the same value to all of them, you don't need to perform separate assignments. Instead, you can chain the assignments together in a single statement, as shown in the code below.

```
x = y = z = 5    # Chained Assignments
print(x, y, z)
----Output----------------
5 5 5
```

Logical Operators

The logical operators help us make decisions based on whether certain things are true or false.

Operator	Operation	Example	Result	Read as
and	Both conditions must be true	True and False	False	True and False (both conditions must be True)
or	Either condition can be true	True or False	True	True or False (one condition must be True)
not	Inverts the condition	not True	False	Not True (inverts True to False)

They are like special tools that help us combine conditions or questions. Let's understand this in more detail with the help of an example. In our previous scenario, where Amelia is planning a trip with Henry and Miles, suppose there are two conditions that must both be met for her to go on the trip:

Condition 1: Amelia must have finished reading at least 2 books.

Condition 2: Henry and Miles must both join her on the trip.

Python's "and" logical operator can help us combine these conditions by checking if both conditions are true. If either condition is false, the "and" operator will return false. Let's write the code for this. In the example code below, suppose Amelia has read 2 books, and both Henry and Miles have agreed to join her on the trip. As you can see, the result is True.

```
books_read = 2
both_joining = True

# Use the "and" operator to check if both conditions are true
if books_read >= 2 and both_joining:
    print("Amelia can go on the trip.")
else:
    print("Amelia can't go on the trip.")
----Output-----------------
Amelia can go on the trip.
```

Now, let's take another example where Amelia has read only 1 book, as shown in the code below. In this case, the condition `books_read >= 2` will be false, while the other condition, `both_joining`, will be `True`. Since both conditions must be true for the trip to happen, the result will be false. So, unfortunately, Amelia can't go on this trip.

```
books_read = 1
both_joining = True

# Use the "and" operator to check if both conditions are true
if books_read >= 2 and both_joining:
    print("Amelia can go on the trip.")
else:
```

```
    print("Amelia can't go on the trip.")
----Output----------------
Amelia can't go on the trip.
```

For the "or" operator, the final result will be True if either condition is true. So, if the rule was that either of these two conditions should be met (not both), then even if Amelia read only one book, she could still have gone on the trip. The Table below lists all the scenarios for the three logical operators. Review it to see how everything works.

Condition 1 (Amelia reads 2 books)	Condition 2 (Henry and Miles join)	Condition 1 AND Condition 2	Condition 1 OR Condition 2	NOT Condition 1	NOT Condition 2
TRUE	TRUE	TRUE	TRUE	FALSE	FALSE
TRUE	FALSE	FALSE	TRUE	FALSE	TRUE
FALSE	TRUE	FALSE	TRUE	TRUE	FALSE
FALSE	FALSE	FALSE	FALSE	TRUE	TRUE

Logical operator precedence

NOT has the highest precedence followed by AND and OR has the lowest. This means NOT is evaluated first in any expression containing these logical operators. AND is evaluated second, after NOT but before OR. OR has the lowest precedence and is evaluated last. For example, if I write

```
not book_read <= 5 and both_joining and book_read > 1
```

then, it means ((not `book_read` <= 5) and both_joining) or `book_read` > 1. Understanding this precedence is important because it affects how complex logical expressions are evaluated by the Python interpreter.

```
x = 7
y = 9
print(not x > 3 and y > 15 or x + y == 16)
----Output----------------
True
```

For example, in the code above,
- `not (x > 3)` is evaluated first. The result is `not (True)` which means `False`.
- This result is then evaluated with `y > 15` using "and". So, `False and (y > 15)` means `False and False`. This results in `False`.
- Finally, this result from the previous step is evaluated with `x + y == 16` using "or". So, `False or (x + y == 16)` means `False or True`, which evaluates to `True`.

You can use parentheses to override the default operator precedence. The code below shows this, and now you can notice the result is totally different from earlier.

```
x = 7
y = 9
print((not x > 3) and (y > 15 or x + y == 16))
----Output----------------
False
```

Review Questions

Read the questions carefully and select the correct options to assess how much you have learned so far. Note that more than one answer may be correct.

2. Which part of `(7 + 4) * 3 / 5` evaluates first?

    a. 3 / 5

    b. 7 + 4

    c. 11 * 3

    d. 4 * 3

3. What is the result of this expression `not (True and False) or (False and not True)`?

    a. False

    b. Error

    c. None

    d. True

4. What is the output of the following code

    ```
    x = 10
    y = 3
    z = x // y * 2 + x % y
    print(z)
    ```

    a. 8

    b. 10

    c. 7

    d. 11

5. Which of the following expressions will result in a floating-point number?
   a. 10 / 5
   b. 10 // 5
   c. 10 * 5
   d. 10 % 5

6. Which operator is used to check if two values are equal in Python?
   a. =
   b. ==
   c. ===
   d. !=

7. What is the difference between the "=" and "==" operators in Python?
   a. "=" is used for comparison, and "==" is used for assigning values.
   b. "=" is used for assigning values, and "==" is used for comparison.
   c. Both are used for assigning values.
   d. Both are used for comparison.

8. What will be the output of the following code?

```
x = 3
y = 3
print(x == y and y > 2 or x < 5)
```

   a. True
   b. False
   c. 0
   d. 1

9. Which of the following statements is true regarding logical operators in Python?
   a. The "or" operator will return True if at least one of the conditions is True.
   b. The "and" operator will return True only if both conditions are True.
   c. The "not" operator negates a condition, turning True into False and vice versa.
   d. "and" and "or" operators have the same precedence in Python.

10. Which of the following is a valid chained assignment in Python?
    a. x, y = 5, 5
    b. x = y = z = 10
    c. x = 5, y = 6
    d. x = y = z

Answer Keys:
1. b; 2. d; 3. c; 4. a; 5. b; 6. b; 7. a; 8. a, b, c; 9. b

8

Numbers

Mathematics with Python

Numbers are a fundamental part of any programming language, and Python is no exception. While coding in Python, you'll frequently work with numbers. You don't need to be an expert mathematician, but it's important to have a basic understanding of the different types of numbers and common operations. In this chapter, we'll focus on two main types of numbers in Python: Integers and Floating-point numbers.

Mathematical Expressions

Before we get into the details of numbers, let's first understand a few basic terms that we will be frequently using throughout. Mathematical Expression is one such term. A mathematical expression is a statement that consists of numbers, variables, operators, and functions. These expressions can be evaluated and simplified to produce a value. Here are some examples of mathematical expressions:

- `x = 2 + 3`
  This can be evaluated to `x = 5`.
- `y = x * (3 + x)`
  For a given value of x (say `x = 2`), this expression can be evaluated to `y = 10`.

In Python, you need to be very explicit when writing an expression. Every operation must be explicitly defined. For example, if you write something like:

```
x = 2(7 + 5)
```

Even though this might look correct from a mathematical perspective, it will throw an error in Python.

```
/Users/anandpandey/mu_code/test1.py:1: SyntaxWarning: 'int' object is
not callable; perhaps you missed a comma?
  x = 2(7 + 5)
Traceback (most recent call last):
  File "/Users/anandpandey/mu_code/test1.py", line 1, in <module>
    x = 2(7 + 5)
TypeError: 'int' object is not callable
```

Python does not automatically assume multiplication. To fix this, you need to explicitly include the multiplication operator *:

```
x = 2 * (7 + 5)
```

Numbers - Integers and Floats

As we discussed in the chapter on Data Types, integers are whole numbers—whether negative, positive, or zero—that do not have a decimal point. In Python, these are referred to as "int." On the other hand, floating-point numbers (called "float" in Python) are numbers that contain a decimal point.

To store int or float values in memory, we use an assignment operator, as shown below:

```
x = 3
y = 4.56
```

Note that we did not use quotation marks, such as "3" or "4.56", to store the values in x and y. Instead, we directly assigned the values. Why? Because these are numbers, not strings.

Converting Numbers

Sometimes, while coding, you might need to convert a float to an int and vice versa. Python has built-in functions to help you with that.

int()

`int()` converts a floating-point number into an integer. As shown in the code below, while converting the float to an integer, the decimal part is truncated.

```
floating_point_number = 23.67
converted_integer = int(floating_point_number)
print(converted_integer)
----Output-----------------
23
```

float()

`float()` converts an integer into a floating-point number. As shown in the code below, a decimal point is added to the integer to make it float.

```
integer_number = 12
converted_float  = float(integer_number)
print(converted_float)
----Output-----------------
12.0
```

What would happen if you divided an integer by another integer?

```
integer_1 = 3
integer_2 = 5
result  = integer_1 / integer_2
print(result)
----Output-----------------
0.6
```

Basic Arithmetic Operators

Python supports the four basic arithmetic operators that you are already familiar with from your mathematics classes in school. The operators are:

1. Addition (+)

    ```
    num_1 = 7
    num_2 = 4
    result = num_1 + num_2
    print(result)
    ----Output----------------
    11
    ```

2. Subtraction (-)

    ```
    num_1 = 7
    num_2 = 4
    result = num_1 - num_2
    print(result)
    ----Output----------------
    3
    ```

3. Multiplication (*)

    ```
    num_1 = 7
    num_2 = 4
    result = num_1 * num_2
    print(result)
    ----Output----------------
    28
    ```

4. Division (/)

```
num_1 = 7
num_2 = 4
result  = num_1 / num_2
print(result)
----Output----------------
1.75
```

Observe the result in the above code. There are some interesting facts here and we will discuss each of them:

- When using a combination of the +, -, and * arithmetic operators, if both operands are integers (int), the result will also be an integer.
- For addition and subtraction, if either operand is a float, the result will be a float, even if the other operand is an integer.

```
num_1 = 7    # Integer
num_2 = 4.5  # Float
result  = num_1 + num_2
print(result)
----Output----------------
11.5         # Float
```

```
num_1 = 7    # Integer
num_2 = 4.5  # Float
result  = num_1 - num_2
print(result)
----Output----------------
2.5          # Float
```

- If the expression includes the division operator (/), the result will always be a float, regardless of the operand types.

```
num_1 = 7     # Integer
num_2 = 4     # Integer
result  = num_1 / num_2
print(result)
----Output-----------------
1.75          # Float
```

- Precedence: Python follows the same order of operations that you have already learned in your math class.

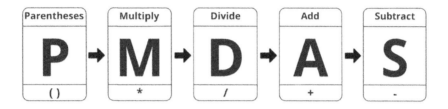

Let's look at the below example to see how Python takes care of these precedence for basic operators.

```
precedence_result = (10 + 5) * 2 - 8 / 4
print(precedence_result)
----Output-----------------
28.0
```

STEP 1. Read the given expression: `(10 + 5) * 2 - 8 / 4` and go from left-to-right.

STEP 2. First evaluate the parentheses: `(10 + 5) = 15`

STEP 3. Perform multiplication: `15 * 2 = 30`

STEP 4. Perform division: `8 / 4 = 2.0`

STEP 5. Perform subtraction: `30 - 2.0 = 28.0`

New Operators

In this section we will discuss three new operators and how Python evaluates them:

1. Floor division (//)
2. Exponentiation (**)
3. Modulus (%)

Floor and Ceiling

Before we dive into these operators, let's first explore the interesting concepts of **floor** and **ceiling** of a number.

- The floor of a number is the greatest integer less than or equal to that number.
- The ceiling of a number is the smallest integer that is greater than or equal to that number.

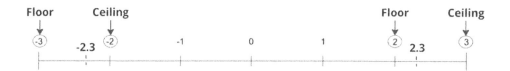

As you can see from the number line above, if we need to calculate the floor and ceiling of the number 2.3:

- The floor will be 2, because 2 is the largest integer that is less than or equal to 2.3.
- The ceiling will be 3, because 3 is the smallest integer that is greater than or equal to 2.3.

Now, if we look at the negative number -2.3

- The floor of -2.3 is -3, because -3 is the largest integer that is less than or equal to -2.3.
- The ceiling will be -2, because -2 is the smallest integer that is greater than or equal to -2.3.

Floor Division

```
x = 7.5
y = 3
division_result = x / y          # Division
floor_division_result = x // y   # Floor Division
print(division_result)
print(floor_division_result)
----Output-----------------
2.5
2.0
```

Python floor division, represented by //, returns the floor of the division (explained in the previous section). The resulting data type depends on the types of the operands:

1. If both operands are integers, the result will be an integer.
   For example, if x = 9 and y = 4, then x // y = 2.
2. If one or both operands are floats, the result will be a rounded-down integer but in float form (with .0 after the decimal point).
   For example, if x = 7.5 and y = 3, then x // y = 2.0.

Exponentiation

Python uses special operator ** for exponentiation which is raising a number to the power of another number. In below figure, 2 is the base which is being raised to the power of 3. So, it will be represented as 2 ** 3 resulting in 8.

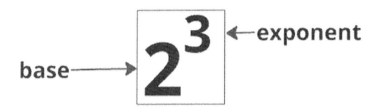

```
x = 2
y = 3
exponentiation_result = x ** y    # Exponentiation
print(exponentiation_result)
----Output----------------
8
```

Operator precedence for **

The exponentiation has higher precedence than most other operators as shown below.

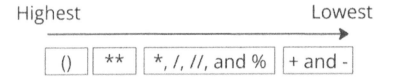

However, there is one interesting thing. When it comes to determining the precedence of multiple **, Python goes from right-to-left. See the following code to understand.

```
x = 2
y = 3
z = 2
exponentiation_result = x ** y ** z        # Multiple Exponentiation
print(exponentiation_result)
----Output-----------------
512
```

It first evaluated `y ** z`, which is $3^2 = 9$. Then, it evaluated `x ** 9`, which is $2^9 = 512$.

Modulo

If we recall our math lessons, we likely learned how to divide two numbers using long division. After the division, we may be left with a remainder, which is the leftover part of the original number. This remainder is very useful in many computer applications. In Python, the modulo operator, represented by the percent symbol (%), is used to find that remainder.

Let's walk through an example where the dividend is 56 and the divisor is 3.

As you can see, this long division results in a quotient of 18 and a remainder of 2.

In programming, you'll often need to calculate this remainder. One common scenario is determining whether a number is odd or even. If the remainder is 0, the number is even; otherwise, it is odd.

Let's take a look at the code below, where we directly calculate the remainder of the division.

```
x = 56
y = 3
modulo_result = x % y        # Modulo / Remainder
print(modulo_result)
----Output----------------
2
```

Random Numbers

Since we're talking about numbers, let's explore an important concept: generating random numbers. This is something you'll often use in game development, simulations, and many other programs where you need a bit of randomness.

Imagine I ask you to give me a random number between 1 and 10. One way to do this would be to write the numbers 1 through 10 on separate pieces of paper, shuffle them, and pick one at random. But what if I asked for a random number between 1 and 10,000? That would be quite a challenge, right? Well, Python can do this in a fraction of a second.

Python has a built-in module called random that allows you to generate random numbers easily. If you're not familiar with the term module, don't worry—we haven't covered that yet. For now, just think of a module as a program file that contains pre-written Python code, which you can reuse in your own programs. All you need to do is write the line "import random" at the beginning of your code to bring in the functionality of the random module.

Now, let's write a code that gives a random integer between 1 and 10, 000.

```
import random
random_num = random.randint(1, 10000)
print(random_num)
```

```
----Output----------------
9893
```

Run the above code again. Every time you run it; you should get a different number.

But what if we want to generate a random float between 1 and 10, 000? Python's random module has another function called **uniform()** that can handle this. To generate random floats, you'll use `random.uniform(a, b)`. This function generates a random float between a and b, **inclusive** of both bounds.

For example, if you wanted a random float between 1 and 10,000, you would write:

```
import random
random_float = random.uniform(1, 10000)
print(random_float)
----Output----------------
4531.29062224415
```

The `random.random()` function generates a random float between 0.0 and 1.0, **exclusive** of both bounds.

```
import random
random_num = random.random()
print(random_num)
----Output----------------
0.6827112137506313
```

Activity: Let's do an activity! We'll write a Python program to create a game where 3 players each roll 2 dice five times. The player with the highest cumulative score at the end wins.

You can use below hints to write this game:

1. Use "import random" at the top of your program.

   ```
   import random
   ```

2. To simulate rolling two dice, you can use `random.randint(1, 6)` twice (once for each die).

   ```
   roll = random.randint(1, 6) + random.randint(1, 6)
   ```

3. For tracking each player's score, create separate variables for each player.

   ```
   player1_score += roll
   ```

Review Questions

Read the questions carefully and select the correct options to assess how much you have learned so far. Note that more than one answer may be correct.

1. Which of the following are valid mathematical expressions in Python?
    a. z = 4(5 + 2)
    b. z = 4 * (5 + 2)
    c. z = 4 + (5 2)
    d. z = (4 + 5) * 2

2. What will be the output of the following code?

    ```
    num = int(19.99)
    print(num)
    ```

    a. 19.99
    b. 19
    c. 20
    d. Error

3. Which of the following statements about division in Python is correct?
    a. / always returns a float, even if both operands are integers.
    b. / always returns an integer if both operands are integers.
    c. // returns an integer if both operands are integers.
    d. // always returns an integer, regardless of the operand types.

4. What will be the output of the following code?

    ```
    num = float(15)
    print(num)
    ```

a. 15
   b. 0.15
   c. Error
   d. 15.0

5. Given the following code, what will be the value of `result`?

   ```
   result = 3 ** 2 ** 2
   print(result)
   ```

   a. 9
   b. 80
   c. 81
   d. 12

6. Which of the following statements about floor division (//) is true?
   a. It returns the largest integer less than or equal to the division result.
   b. It rounds the division result to the nearest whole number.
   c. If one operand is a float, the result will still be a float but rounded down.
   d. It performs regular division and discards the remainder.

Answer Keys:
1. b and d; 2. b; 3. a and c; 4. d; 5. c; 6. a

9

Strings

Magic with Text

In the previous chapter, we learned that a string is a data type. We also discovered that strings are essentially sequences of characters. Think of characters as beads and the string as a necklace made up of those beads. These characters can be letters, numbers, symbols, or even spaces. Strings are enclosed in either single quotes (' ') or double quotes (" ").

In this chapter, we will focus on the different operations we can perform on strings.

String Creation

The first question that comes to mind is: How do I create a string? Well, as we learned in the chapter on data types, strings are sequences of characters enclosed in either single or double quotes. In fact, there are several ways to create a string in Python, and in the code below, you'll see four common methods:

1. Using single quote
2. Using double quotes
3. Using triple single quotes
4. Using triple double quotes

```
name1 = 'Amelia'
name2 = "Amelia"
name3 = '''Amelia
Smith in two lines'''
name4 = """Amelia
Smith in two lines"""
```

Let us look at ways we can handle situations where you need to use quotes or quotations marks as part of the string.

- Using Different Quotes for the String and Inside the String.
    You can use single quotes for the string and double quotes inside it, or vice versa. This way, Python can distinguish between the quotes used to define the string and the ones that are part of the string's content.

    ```
    example_1 = 'He said, "I know Python Programming!"'
    print(example_1)
    example_2 = "Let's create something fun with Python."
    print(example_2)
    ----Output----------------
    ```

```
He said, "I know Python Programming!"
Let's create something fun with Python.
```

- Using a Backslash (\).

    If you want to use the same type of quote inside the string as the one you used to define the string, you can escape the inner quote using a backslash (\).

    ```
    example_1 = "He said, \"I know Python Programming!\""
    print(example_1)
    example_2 = 'Let\'s create something fun with Python.'
    print(example_2)
    ----Output----------------
    He said, "I know Python Programming!"
    Let's create something fun with Python.
    ```

- Using Triple Quotes.

    For strings that contain both single and double quotes, or when you need a multiline string, you can use triple quotes (''' or """). This method is especially useful when you are working with strings that include multiple quotes or require complex formatting.

    ```
    example_1 = '''He said, "I know Python Programming!"'''
    print(example_1)
    example_2 = """Let's create something fun with Python."""
    print(example_2)
    ----Output----------------
    He said, "I know Python Programming!"
    Let's create something fun with Python.
    ```

Would the following code work?

```
name1 = """
```

Why or why not?

No, this will not work. Triple quotes in Python (`"""` or `'''`) are used to define multi-line strings. If you only provide an opening triple quote (`"""`) without a corresponding closing triple quote, Python will raise a syntax error because it expects the string to be properly enclosed.

String Concatenation

String concatenation means joining two or more strings sequentially to form a larger string. It's like combining words to create a sentence. To achieve this, you use the + operator. The + operator acts like glue, sticking the strings together. Look at the example code below. Did you notice that the two strings, "Amelia" and "Smith", are joined without any spaces in between? That's because the `+` operator does not add anything - it simply combines the strings as they are.

```
first_name = "Amelia"
last_name = "Smith"
print(first_name + last_name)
----Output----------------
AmeliaSmith
```

How can you add space between Amelia's first and last name? You can do this by using an additional "+" to add a space, as shown in the code below. Another way to do it is by passing `first_name` and `last_name` as two separate arguments to the `print()` function.

```
first_name = "Amelia"
```

```
last_name = "Smith"
print(first_name + " " + last_name)
print(first_name, last_name)
----Output-----------------
Amelia Smith
Amelia Smith
```

Length of a String

Python has a built-in function **len()** that let you find the total number of characters in a string including letters, numbers, spaces, and symbols. This is called length of a string.

You need to pass the string as the argument to this function and it returns the length of the passed string.

```
sentence = "I'd get 2 candies."
length_of_sentence = len(sentence)
print(length_of_sentence)
----Output-----------------
18
```

The characters in the sentence variable are – `I`, `'`, `d`, `space`, `g`, `e`, `t`, `space`, `2`, `space`, `c`, `a`, `n`, `d`, `i`, `e`, `s`, `.`
If you count all these characters, you will get a total of 18. That is exactly what the `len()` function did. It counted them for you. So, `len()` is very helpful when you need to know the length of a string or count how many characters you have.

Repetition

If we ever need to repeat a string a certain number of times, we can do so easily using the asterisk (*) operator, as shown below.

```
name = "Amelia"
repeated_name = name * 3
print(repeated_name)
----Output----------------
AmeliaAmeliaAmelia
```

Here, the string `"Amelia"`, which is assigned to a variable `name`, is repeated 3 times, resulting in `AmeliaAmeliaAmelia`.

Accessing Characters in a String (Indexing)

If I asked you to write a program where you ask the user for their name and display the third character of the name, how would you do that? Python makes it easy to access each character of a string. It assigns a position number to each character within the string, which allows you to access any character by its position number. This position number is called the index. The below figure shows the indexing for the string "Amelia".

String →	A	M	E	L	I	A
Index →	0	1	2	3	4	5

Notice that the first character of a string starts at position 0. Can you figure out the position of the last character without manually calculating it? You can use the `len()` built-in function, which we learned earlier. The `len()` function returns the length of the string. If we subtract 1 from the length, we get the position of the last character, as shown below.

```
name = "Amelia"
last_char_position = len(name) - 1
```

```
print(last_char_position)
----Output----------------
5
```

Now, let's see how we can access a character of a string using Python indexing. To access a character, we can suffix the string variable with [index], where index is the position of the character. For example, to access the third character of the variable `name` in the below code, we use `name[2]`. We use 2 (and not 3) because indexing starts at 0, so the third character is at index 2.

```
name = "Amelia"
third_character = name[2]
print(third_character)
----Output----------------
e
```

Interestingly, there is another way to access the last character of a string. What happens if we use a negative number as the index? When you use a negative index, it starts counting from the end of the string. So, the last character of any string is at index -1.

String	A	M	E	L	I	A
Index	0	1	2	3	4	5
	-6	-5	-4	-3	-2	-1

Following code shows how we can access the last character using the negative indexing.

```
name = "Amelia"
last_character = name[-1]
print(last_character)
----Output----------------
a
```

Slicing Strings

Just like slicing a pizza or cake, you can also slice strings to get a smaller string, called a substring. To do this in your code, you need to know where to start slicing and where to end. You slice a string by appending `[start_index : end_index]` to the string. The `start_index` is "**inclusive**," meaning the character at that index will be included in the result. The `end_index` is "**exclusive**," meaning the character at that index will not be included in the result. If you don't provide a `start_index`, Python assumes you want to start from the beginning (index 0). Similarly, if you don't provide an `end_index`, Python assumes you want to go all the way to the end of the string. Confused? The best way to understand this is to look at the example code below and try it out yourself in your Mu Editor. It will become clear once you experiment with it. Notice how the behavior changes when you leave out either the start or end index.

```
name = "Amelia Smith"
print(name[0:3])
print(name[2:])
print(name[:4])
----Output----------------
Ame
elia Smith
Amel
```

In the example shown above, note the following:
- `name[0:3]` gives the part of the string from index `0` to index `2` (not including index `3`), so it gives `Ame`.
- `name[2:]` gives the part of the string starting from index `2` to the end, which is `elia Smith`.
- `name[:4]` gives the part from the start of the string up to index `3`, which is `Amel`.

What will `name[1:4]` give? Try it out. It should give `mel`.

String Formatting

Sometimes, you may need to include variables inside a string. You can embed variables directly into string by using formatted strings (or f-strings). This is done by prefixing the string with "f" or "F". F-strings are especially helpful for more complex sentences. Look at the example shown below.

```
first_name = "Amelia"
age = 13
formatted_text = f"{first_name} is {age} years old."
print(formatted_text)
----Output----------------
Amelia is 13 years old.
```

The **{}** are placeholders for variables, and the **f** before the string tells Python to fill in those placeholders with the actual values of `first_name` and age.

String Methods

Methods are a type of function, but they belong to a specific class. We'll learn more about classes later, but for now, just know that a string is a class in Python that has many useful functions built into it that can help you manipulate and work with them more easily. Because these functions are specific to strings, we call them "methods." These methods are like special tools for handling strings. To use them, you call the method by appending its name to the string, separated by a dot (`.`). Let's take a look at some of the useful methods in the examples below.

Changing Case (Uppercase / Lowercase)

What if you want to change the case of a string? For example, you have a variable name with the value "`amelia`," but you want to display it as `AMELIA` or the other way around. Python provides two built-in methods to help you with this: `.upper()` and `.lower()`, as shown below.

```
name = "Amelia Smith"
print(name.upper())
print(name.lower())
----Output----------------
AMELIA SMITH
amelia smith
```

- `name.upper()` converted the string "`Amelia Smith`" to "`AMELIA SMITH`" (all upper case).
- `name.lower()` converted the string "`Amelia Smith`" to "`amelia smith`" (all lowercase).

Capitalize

If you ever want to convert the first character of a string to uppercase and all other characters to lowercase, you can use the `.capitalize()` method.

```
name = "amElia SMITH"
print(name.capitalize())
```

```
----Output----------------
Amelia smith
```

As shown in the code below, the first name was originally in all lowercase and last name was all in uppercase, but the .capitalize() method changed the first character to uppercase and kept the rest in lowercase.

Title

The `title()` method is used to capitalize the first letter of each word in a string. It's like when you write the title of a book, where each important word starts with a capital letter.

```
name = "amelia smith"
print(name.title())
----Output----------------
Amelia Smith
```

If you have a sentence and want to make sure each word starts with a capital letter, `title()` is really helpful. It ensures proper capitalization.

Strip

This method removes any extra spaces from the beginning and end of a string.

```
text = "      Amelia         "
print(text.strip())
----Output----------------
Amelia
```

Replace

This method allows you to replace a specific part of the string with something else. This replaces all occurrences.

```
text = "My name is Amelia"
print(text.replace("Amelia", "Henry"))
----Output-----------------
My name is Henry
```

Find

This method tells you the position (or index) where a certain letter or word appears in the string. This returns the first occurrence of the text.

```
text = "My name is Amelia"
print(text.find("Amelia"))
----Output-----------------
11
```

In the example shown above, count the position where the word `"Amelia"` starts. What do you get? You should get `11`. Remember, counting starts from `0`.

There are several other methods that strings offer, but the ones above are the most frequently used. If we need to explore other methods in later chapters, we'll learn about them then.

Chapter 9

Review Questions

Read the questions carefully and select the correct options to assess how much you have learned so far. Note that more than one answer may be correct.

1. Which operator is used to concatenate two strings in Python?
    - a) &
    - b) *
    - c) +
    - d) //

2. What is the result of `len("I've got 2 candies.")`?
    - a) 16
    - b) 17
    - c) 18
    - d) 19

3. What index would you use to access the fourth character of the string `"Miles"`?
    - a) 3
    - b) 4
    - c) 5
    - d) 0

4. What will `name[2:]` return if `name = "Amelia Smith"`?
    - a) Amelia
    - b) Amelia Smith
    - c) Amel

d) `elia Smith`

5. What will code `print("Amelia"[:])` print?
   a) Nothing
   b) Amelia
   c) meli
   d) Error

Answer Keys:
1. c; 2. d; 3. a; 4. d; 5. b

10

List, Tuples, and Dictionaries

Sequences and Mappings

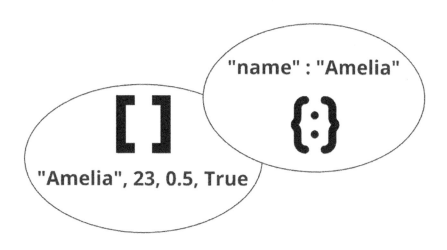

Imagine you have a bunch of sports gear for different sports, like soccer, tennis, and basketball. Now, someone asks you to quickly find the soccer boots. How can you do it efficiently? Well, it all depends on how you organize your sports gear.

Option 1: Organizing by Sequence

You could arrange your gear in different boxes (one for soccer, one for tennis, one for basketball) and then arrange those boxes in a sequence (like 1st box for soccer, 2nd box for tennis, etc.).

How this helps: If I ask for the soccer boots, you can immediately think, "Oh, soccer gear is in the 1st box!" and quickly grab it. This method relies on order.

This is like a Python list or tuple that you can use to store your items in a specific order and retrieve them using their position (index) in the list.

Option 2: Organizing with Labels (Keys)

In another way, you could keep different sports gears in different boxes but this time, you can label the boxes with some specific names or "keys". So, you could put a sticky note that says "Soccer Gear" on the soccer gear box.

How this helps: If I ask for soccer boots, you'll just look at the "Soccer Gear" label and immediately grab the boots from that box, without needing to remember which position it's in. You're using labels (or keys) to find what you need!

This is like a Python dictionary that you can use to store key-value pairs. Each "key" (like "Soccer Gear") is used to find a corresponding "value" (like the actual gear inside the box). The key helps you look up the item quickly.

Now that we clearly understand what sequences and mappings are, let's take a deeper dive into lists, tuples, and dictionaries to explore them further.

Data Structure

As you saw in the examples above, we need to organize our data efficiently to retrieve it quickly. Similarly, Python uses data structures to organize items in the computer's memory, making it faster to retrieve and modify them. Lists, tuples, and dictionaries are all examples of Python's data structures. You'll often encounter terms like mutable and immutable when working with data structures. **Mutable** means something that can be changed after it's created, while **immutable** refers to items that cannot be changed once created. In Python, some data structures, like lists and dictionaries, are mutable, whereas others, like tuples, are immutable. I wanted to clarify these terms since we'll be using them while discussing lists, tuples, and dictionaries.

List

A list allows you to store many values in a single variable, and it keeps track of these values using something called an **index**. Let's understand this with an example.

Suppose you have a sequence of seven values:

```
8, 7, 12, 56, 2, 0, -9
```

If you want to store these values and be able to retrieve them, you might think of creating seven separate variables and assigning each value to one:

```
number_1 = 8
number_2 = 7
number_3 = 12
number_4 = 56
number_5 = 2
number_6 = 0
number_7 = -9
```

But what if I tell you there's a more efficient way to handle this? Instead of creating separate variables, you can use a list to store all these values in a single variable.

This is how you can organize it:

numbers	
index	value
0	8
1	7
2	12
3	56
4	2
5	0
6	-9

In Python, you can store this sequence of numbers in a list:

```
numbers = [8, 7, 12, 56, 2, 0, -9]
```

Each value in the list has a **position** called an **index**, starting from 0. So, if I ask you to give me the **4th value** (which is 56), you can simply refer to the list and access the value at **index 3** (because counting starts from 0). This means:

```
print(numbers[3])
----Output----------------
56
```

In this way, with just **one variable** (the list), you can store all your values and easily access any of them by their **index**.

Creating a List

The simplest way to create a list is to use an empty set of square brackets `[]` to represent an empty list and then store that in a variable using an assignment statement. If I want to create a list say `list_1`, then I would do the following:

```
list_1 = []
```

This will create an empty list. If I want to create a list to store a string "Hello" and a number 9 then I can do something like

```
list_1 = ["Hello",9]
print(list_1)
----Output----------------
['Hello', 9]
```

Another way to create a list is by using the built-in `list()` function (also known as a **constructor** in object-oriented programming, which we'll explore later in this book). This method is especially useful when you want to convert other objects, such as strings or tuples, into a list.

```
list_1 = list([9, -2, 13, "Hello", 5])
print(list_1)
----Output----------------
[9, -2, 13, 'Hello', 5]
```

Let's summarize the key characteristics of a list.

1. Lists maintain the order of elements. The position of each element is defined by its index, starting from 0.
2. You can change, add, or remove elements after the list is created. So, the lists are mutable.

```
numbers = [8, 7, 12, 56, 2, 0, -9]
print(numbers[2])
numbers[2] = 99
print(numbers[2])
print(numbers)
----Output----------------
12
99
[8, 7, 99, 56, 2, 0, -9]
```

3. Each element in a list has a specific index.
4. Lists can contain duplicate values.

    So, you can have a list number with values [7, 7].

```
numbers = [7,7]
print(numbers[0])
print(numbers[1])
----Output----------------
7
7
```

5. Lists can hold elements of different data types, such as integers, strings, and even other lists.

```
item_1 = 10
item_2 = -3.9
item_3 = "Hello, World!"
item_4 = [3, 7, 9]
```

```
list_1 = [item_1, item_2, item_3, item_4]
print(list_1)
----Output----------------
[10, -3.9, 'Hello, World!', [3, 7, 9]]
```

Isn't it cool that a list can hold another list?

Nested List

A nested list in Python is simply a list within a list. This means that one or more elements of a list can be other lists themselves.

Creating a nested list

You can create a nested list by including lists as elements within another list.

```
list_1 = [[8, 7, 12], [56, 2, 0], [-9, 4, 5]]
```

As you can see, there are three lists nested within `list_1`. The last item of `list_1` is `[-9, 4, 5]`. What if you want to access the second value (which is `4`) of this last item in `list_1`?

You can access it using nested indexing:

```
print(list_1[2][1])
----Output----------------
4
```

Interestingly, you can also do a reverse indexing like below:

```
print(list_1[-1][1])
print(list_1[-1][-2])
```

```
----Output----------------
4
4
```

Accessing List Elements

Items in a list are also called elements. You can access these list elements using their index. Indexing starts from `0` for the first item, `1` for the second item, and so on.

For example, in the list `my_list = [10, 20, 30]`:

- `my_list[0]` will give you `10` (the first element)
- `my_list[1]` will give you `20` (the second element)

You can also use negative indexing, which allows you to access elements starting from the end of the list:

- `my_list[-1]` will give you `30` (the last element)
- `my_list[-2]` will give you `20` (the second-to-last element)

This reverse indexing works similarly to accessing characters in a string, which you've already learned in the previous chapter.

Slicing a List

Just like you can slice a cake, you can also slice a list! And guess what you get? A portion of the list. This works very similarly to string slicing.

The syntax for slicing a list is:

```
list[start:end]
```
- start: The index where the slice begins (inclusive).
- end: The index where the slice ends (exclusive), meaning the element at the end index is not included.

Let's see an example:

```
names = ["Amelia", "Henry", "Smith", "Miles", "Lila"]
print(names[0:3])
print(names[2:])
print(names[:4])
----Output----------------
['Amelia', 'Henry', 'Smith']
['Smith', 'Miles', 'Lila']
['Amelia', 'Henry', 'Smith', 'Miles']
```

In the example shown in the above code, note the following:
- `names[0:3]` gives the part of the string from index 0 to index 2 (not including index 3).
- `names[2:]` gives the part of the string starting from index 2 to the end.
- `names[:4]` gives the part from the start of the string up to index 3.

What will `names[1:4]` give? Try it out. It should give `['Henry', 'Smith', 'Miles']`.

Adding Elements to a List

Sometimes, you may want to create an empty list first and then add more items to it later. Also, you might want to create a list with a few items initially and then expand it by adding another item later. You can do this using the **append()** method.

Creating an empty list and adding items later

```
list_1 = []
print(list_1)
list_1.append("Hello")
print(list_1)
----Output----------------
[]
['Hello']
```

Starting with a list and expanding it

```
list_1 = [-23.5, 10]
print(list_1)
list_1.append("Hello")
print(list_1)
----Output----------------
[-23.5, 10]
[-23.5, 10, 'Hello']
```

Inserting an element

You can insert an element at a specific index in a list using the **insert()** method. This allows you to add an item at any position in the list.

In the example below, we are inserting the color `"green"` after index `1`, which means it will be placed at index `2`:

```
colors = ["red", "pink", "blue"]
colors.insert(2, "green")
print(colors)
----Output----------------
['red', 'pink', 'green', 'blue']
```

Notice that when you inserted an element at index 2, the other items in the list shifted to the right to make space for the new element. This is how lists in Python automatically adjust when you insert an element at a specific position.

Extending a list

You can extend a list by adding multiple elements to the end of it using the **extend()** method. This is especially useful when you have one list, and you want to add all its elements to the end of another list.

```
colors_1 = ["red", "pink", "blue"]
colors_2 = ["purple", "green"]
colors_1.extend(colors_2)
print(colors_1)
----Output----------------
['red', 'pink', 'blue', 'purple', 'green']
```

Notice that the entire contents of `colors_2` are added to `colors_1` without creating a nested list.

Remove Elements from a List

You can remove the first occurrence of an element from a list by its value using the **remove()** method. Unlike other methods we've discussed so far, we don't need to specify the index, just the value of the element you want to remove.

```
colors = ['red', 'pink', 'blue', 'purple', 'pink']
colors.remove('pink')
print(colors)
----Output----------------
['red', 'blue', 'purple', 'pink']
```

Notice that we tried removing "pink", and since there were two occurrences of "`pink`" in the list, it only removed the first one.

Popping an Element from a List

Now, imagine you need to remove an element from a list, not by its value, but by its index. Additionally, you want to retrieve the removed item so you can use it in your code. Python provides a built-in method called `pop()` to do this.

The `pop()` method removes an element at a specified index and returns it. If no index is provided, it removes and returns the last item of the list.

```
colors = ['red', 'pink', 'blue', 'purple', 'pink']
removed_color = colors.pop(2)
print(colors)
print(removed_color)
last_color = colors.pop()
print(colors)
print(last_color)
----Output----------------
['red', 'pink', 'purple', 'pink']
blue
['red', 'pink', 'purple']
pink
```

Let's understand what this code is doing:
1. The `pop(2)` method removes the element at index 2 (which is `'blue'`) and returns it.
2. The list now becomes: `['red', 'pink', 'purple', 'pink']`
3. The color that was removed at index 2 is `'blue'`. So, the value of `removed_color` is 'blue'.

4. The `pop()` method, when called without an index, removes the last element in the list, which is 'pink' (the last occurrence).
5. The list now becomes: `['red', 'pink', 'purple']`
6. The color that was removed at the last index is 'pink'. So, the value of `last_color` is 'pink'.

One important thing to notice here is that the `pop()` method both modifies the list and returns the removed item, allowing you to use it in your code.

Clearing a List

If you want to remove all elements from a list and get an empty list, you can use the **clear()** method. This is very useful when you need to reset a list and start fresh without creating a new list.

```
colors = ['red', 'pink', 'blue', 'purple', 'pink']
colors.clear()
print(colors)
----Output----------------
[]
```

Modifying a List

It's possible to change an element of a list by directly accessing it through its index. For example, if you want to change the color at index 2 (which is "blue") to another color, such as "green", you can do it like this:

```
colors = ['red', 'pink', 'blue', 'purple', 'pink']
colors[2] = "green"
print(colors)
----Output----------------
['red', 'pink', 'green', 'purple', 'pink']
```

Finding Elements

If you want to find the **first occurrence** of an item within a list, you can use **index()** method.

```
colors = ['red', 'pink', 'blue', 'purple', 'pink']
print(colors.index('pink'))
----Output----------------
1
```

The `index()` method can also accept additional arguments, following the below syntax:

```
list.index(item, start, end)
```

The `index()` method in Python has the following parameters:

- item (required): This is the element you are searching for. For example, in the previous example, we searched for the item 'pink'.
- start (optional): This is the index where the search will begin. If not provided, the search starts from the beginning of the list.
- end (optional): This is the index where the search will end. If not provided, the search goes till the end of the list.

How will you find the 2nd occurrence? Try yourself first and then check you answer with the following approach:

To find the second occurrence of an item in a list, you can use a combination of list methods. One way to do this is by using the `index()` method to find the first occurrence and then search for the next one starting from the position right after the first occurrence.

```
colors = ['red', 'pink', 'blue', 'purple', 'pink']
first_index = colors.index("pink")
second_index = colors.index("pink", first_index + 1)
print(second_index)
----Output-----------------
4
```

Total Number of Occurrences

You can use the count() method to find the total number of occurrences of an item within a list.

```
colors = ['red', 'pink', 'blue', 'purple', 'pink']
print(colors.count('pink'))
----Output-----------------
2
```

Length of a List

The length of a list is the total number of items contained in that list.

```
colors = ['red', 'pink', 'blue', 'purple', 'pink']
print(len(colors))
----Output-----------------
5
```

List Concatenation

In the chapter on strings, we learned that the + operator concatenates two strings to create a combined string. You can use + with lists as well. It will combine the lists, creating a new list that contains all the items from both lists.

```
colors1 = ['red', 'pink', 'blue', 'purple', 'pink']
colors2 = ['green', 'magenta']
combined_colors =colors1 + colors2
print(combined_colors)
----Output----------------
['red', 'pink', 'blue', 'purple', 'pink', 'green', 'magenta']
```

Tuple

Tuples are very similar to lists, so you might wonder why we even need to learn about them. The key difference is that lists are mutable, meaning you can change, add, or remove elements. On the other hand, tuples are immutable. This means once a tuple is created, you cannot change, add, or remove its elements.

Why would we need something like this? In programming, there are situations where you want to store data but protect it from being modified during the execution of your program. A tuple provides a way to ensure that the data remains unchanged, offering an added layer of safety.

Creating Tuples

You can create a tuple by placing items inside parentheses "()" and separating them by commas.

```
colors = ('red', 'pink', 'blue', 'purple', 'pink')
```

```
print(type(colors))
print(colors)
----Output----------------
<class 'tuple'>
('red', 'pink', 'blue', 'purple', 'pink')
```

One interesting point to note is that if you try to create a tuple with only one element, the following code won't work as expected:

```
colors = ('red')
print(type(colors))
print(colors)
----Output----------------
<class 'str'>
red
```

As you can see, it creates a string instead of a tuple. This happens because Python treats ('red') as a string, not a tuple.

So, how do you create a tuple with a single item? The correct way is by adding a comma after the item, like this:

```
colors = ('red',)
print(type(colors))
print(colors)
----Output----------------
<class 'tuple'>
('red',)
```

Supported Methods

Tuples do not support all the methods that lists do, primarily because they are immutable. As a result, methods like `append()`, `insert()`, `remove()`, `pop()`, `clear()`, and `extend()` — which modify the list—will not work with tuples.

However, tuples do support methods that only read their values. Tuples support the `count()` and `index()` methods. You can also access tuple elements by their index, just like you would with a list.

Dictionary

At the start of this chapter, we discussed option-2, which involves organizing items using labels (or keys). Dictionaries allow you to store items by associating them with unique keys. Unlike lists and tuples, dictionaries are unordered, meaning their items are not stored in any specific sequence, so you cannot access them by an index. However, you can access their items by using their keys.

In simple terms, a dictionary is like a phonebook, where you look up a person's name (the key) to find their phone number (the value).

Dictionaries are mutable, meaning you can add, remove, or modify items in a dictionary at any time. Each item in a dictionary consists of a key and its associated value, which is why dictionaries are also referred to as key-value pairs.

Creating a Dictionary

You can create a dictionary in Python by placing key-value pairs inside curly braces `{}`, with each pair separated by a comma. The key and the value are separated by a colon `:`.

Here's an example of how to create a dictionary:

```
person = {
    "name": "Amelia",
    "age": 12,
    "phone": "xxx-xxx-xxxx",
    "address": "123 Python Way"
}
print(type(person))
print(person)
----Output----------------
<class 'dict'>
{'name': 'Amelia', 'age': 12, 'phone': 'xxx-xxx-xxxx', 'address': '123 Python Way'}
```

In the above code example, we can observe the following:
- person is a variable of type dictionary (Python refers to it as '**dict**').
- The dictionary person contains details about Amelia, organized as key-value pairs.

Accessing Dictionary Elements

You can access an element in a dictionary in a similar way as you would for lists and tuples (using square brackets `[]`). However, instead of using an index, you use the key to retrieve the corresponding value.

```
person = {
    "name": "Amelia",
    "age": 12,
    "phone": "xxx-xxx-xxxx",
    "address": "123 Python Way"
}
print(person["name"])
----Output----------------
```

```
Amelia
```

The issue with accessing dictionary elements directly using square brackets `[]` is that you must supply the exact key. If you provide a non-existing or incorrect key, Python will throw a `KeyError`.

However, there is a safer way to access dictionary elements using the `get()` method. The `get()` method will return `None` if you provide a key that doesn't exist in the dictionary, instead of raising an error.

```
person = {
    "name": "Amelia",
    "age": 12,
    "phone": "xxx-xxx-xxxx",
    "address": "123 Python Way"
}
print(person.get("age"))
----Output----------------
12
```

You can also provide a default value to be returned if the key doesn't exist:

```
person = {
    "name": "Amelia",
    "age": 12,
    "phone": "xxx-xxx-xxxx",
    "address": "123 Python Way"
}
print(person.get("email", "Not Available"))
----Output----------------
Not Available
```

As shown above, since `"email"` is not found in the dictionary, the method returned `"Not Available"` instead of `None`.

Adding to Dictionary

You can add a new item to a dictionary by simply assigning a value to a new key. If the key doesn't exist in the dictionary, Python will create it and assign the value.

```
person = {
    "name": "Amelia",
    "age": 12,
    "phone": "xxx-xxx-xxxx",
    "address": "123 Python Way"
}
person["email"] = "amelia@example.com"
print(person)
----Output----------------
{'name': 'Amelia', 'age': 12, 'phone': 'xxx-xxx-xxxx', 'address': '123 Python Way', 'email': 'amelia@example.com'}
```

Modifying a Dictionary

To modify an existing item in a dictionary, you simply assign a new value to an existing key. This will replace the old value with the new one.

```
person = {
    "name": "Amelia",
    "age": 12,
    "phone": "xxx-xxx-xxxx",
    "address": "123 Python Way"
}
person["age"] = 13
print(person)
----Output----------------
```

```
{'name': 'Amelia', 'age': 13, 'phone': 'xxx-xxx-xxxx', 'address': '123 Python Way'}
```

Removing Items from a Dictionary

For removing items from a dictionary, there are several methods, and you can use any of them depending on your needs:

- del

  ```
  del person["age"]
  print(person)
  ----Output----------------
  {'name': 'Amelia', 'phone': 'xxx-xxx-xxxx', 'address': '123 Python Way'}
  ```

- pop()

  This will return the removed value.

  ```
  age = person.pop("age")
  print(age)
  print(person)
  ----Output----------------
  12
  {'name': 'Amelia', 'phone': 'xxx-xxx-xxxx', 'address': '123 Python Way'}
  ```

- clear()

  This is similar to clearing a list that we learned earlier. It will remove all items from the dictionary.

  ```
  person.clear()
  print(person)
  ```

```
----Output----------------
{}
```

Length of a Dictionary

The length of a dictionary is the total number of key-value pairs contained in that dictionary.

```
person = {
    "name": "Amelia",
    "age": 12,
}
print(len(person))
----Output----------------
2
```

Merging Dictionaries

Earlier, we saw that we can merge two lists using either the + operator or the `extend()` method.

- The + operator concatenates two lists, creating a new list.
- The `extend()` method, on the other hand, mutates the first list by adding elements from the second list to it.

For tuples, since they are immutable, there is no extend() method available. The only way to merge tuples is by using the + operator, which creates a new tuple by concatenating them.

In the case of dictionaries, since they are mutable, you can use the **update()** method to merge two dictionaries. The `update()` method will modify the first dictionary by adding the key-value pairs from the second dictionary.

```
person = {
    "name": "Amelia",
    "age": 12,
}
person_more = {
    "phone": "xxx-xxx-xxxx",
    "address": "123 Python Way"
}
person.update(person_more)
print(person)
----Output-----------------
{'name': 'Amelia', 'age': 12, 'phone': 'xxx-xxx-xxxx', 'address': '123 Python Way'}
```

Dictionary Unpacking

Python has an interesting double asterisk (**) operator that, when prefixed with a dictionary, unpacks the dictionary. But what does "unpacking" mean? Unpacking a dictionary means breaking it into individual key-value pairs.

For example, if we write `{**person}`, Python is literally breaking down the dictionary person into its individual key-value pairs and inserting them into a new dictionary.

Now, let's look at how we can merge two dictionaries using the `**` operator:

```
person1 = { "name": "Amelia", "age": 12 }
person2 = { "phone": "xxx", "address": "123" }
merged_person = {**person1, **person2}
```

```
print(merged_person)
----Output----------------
{'name': 'Amelia', 'age': 12, 'phone': 'xxx', 'address': '123'}
```

Let's understand what's going on:
- `person1` contains two key-value pairs: `{"name": "Amelia", "age": 12}`.
- `person2` contains two key-value pairs: `{"phone": "xxx", "address": "123"}`.

When we use `**person1`, it unpacks the dictionary `{"name": "Amelia", "age": 12}`, extracting each key-value pair individually: `"name": "Amelia"` and `"age": 12`. Notice that the curly braces `{}` are no longer present after unpacking; we just get the individual key-value pairs.

Similarly, when `**person2` is unpacked, it extracts `"phone": "xxx"` and `"address": "123"` - again, without the curly braces.

Now, by doing `{**person1, **person2}`, a new dictionary `{}` is created, and the unpacked items from both `person1` and `person2` are placed inside this new dictionary. As a result, the new dictionary becomes: `{"name": "Amelia", "age": 12, "phone": "xxx", "address": "123"}`. This merged dictionary is then assigned to the variable `merged_person`.

Nested Dictionaries

Dictionaries can also contain other dictionaries as values for certain keys. This is known as a nested dictionary. Let's look at an example:

```
person1 = { "name": "Amelia", "age": 12 }
person2 = { "phone": "xxx", "address": "123" }
nested_dict = {
```

```
        "person1" : person1,
        "person2" : person2
}
print(nested_dict)
----Output-----------------
{'person1': {'name': 'Amelia', 'age': 12}, 'person2': {'phone': 'xxx',
'address': '123'}}
```

Review Questions

Read the questions carefully and select the correct options to assess how much you have learned so far. Note that more than one answer may be correct.

1. Which of the following statements about lists, tuples, and dictionaries in Python are TRUE?
    a) Lists are mutable, meaning their elements can be changed after they are created.
    b) Tuples are immutable, meaning their elements cannot be changed once created.
    c) Dictionaries are mutable, but they don't maintain the order of elements.
    d) Lists and dictionaries can hold items of different data types, but tuples can only hold elements of one type.

2. Which of the following methods can be used to remove an item from a list in Python?
    a) `pop()`
    b) `remove()`
    c) `clear()`
    d) `del`
    e) `discard()`

3. Consider the following tuple:

    ```
    t = ('apple', 'banana', 'cherry')
    ```

Which of the following operations are valid?
```
a) t[0] = 'orange'
b) t.append('date')
c) t[1]
d) t.count('banana')
```

4. What will the following code print?

```
nested_dict = {
    "person1": {"name": "Alice", "age": 30},
    "person2": {"name": "Bob", "age": 25}
}
print(nested_dict["person1"]["name"])
```

```
a) Alice
b) Bob
c) Error: Cannot access nested dictionary
d) person1
```

5. If you try to access a non-existing key in a dictionary using [] notation, what will happen?
   a) It will return `None`.
   b) It will raise a `KeyError`
   c) It will raise an `IndexError`
   d) It will return the default value provided by `get()`

Answer Keys:
1. a, b, c; 2. a, b, c, d; 3. c, d; 4. a; 5. b

11

Conditional Statements

Making Decisions Using If-Else

How do we make decisions in real life? Often, we think like this: "If it snows tomorrow, then we won't have school," or "Mom says, if you eat your green vegetables, then you can watch a movie." These are examples of **conditional statements** that help make decisions or choices based on conditions. The action depends on whether a certain condition is true. In the first statement, the condition is "if it snows tomorrow," and the action is "school will be closed." In the second statement, the condition is "if you eat your green vegetables," and the action is "you can watch a movie." The action happens only when the condition is true.

In coding, these conditions are called boolean expressions (which evaluate to either `True` or `False`). We learned about boolean values in the chapter on logical operators. Here are a few examples of boolean expressions or conditions:

- `chances_of_snow > 50%`
- `score_in_the_game < 100`
- `color_of_shirt != 'Red'`

Each of these conditions will evaluate to either `True` or `False`, depending on the situation. For example, if `chances_of_snow` is greater than `50`, the first condition would be `True`; otherwise, it would be False. Similarly, the other conditions will evaluate based on the values provided.

If Statement

The if statement is the most basic conditional. It checks whether a condition is True. If it is, the code inside the if block is executed. If not, the code is skipped.

```
if condition:
    # Code to execute if the condition is true
else:
    # Code to execute if the condition is false
```

Let's write a code to see how we can implement our understanding of conditional statements and observe how the code behaves.

```
color_of_shirt = 'green'
if color_of_shirt == 'red':
    print("Amelia's shirt is red.")
print("Amelia's shirt is not red.")
----Output----------------
Amelia's shirt is not red.
```

If you look at the code above, you can observe the following:
- The condition `color_of_shirt == 'red'` evaluates to `False` because `color_of_shirt = 'green'`.
- Since the condition is `False`, the code inside the if block is not executed and is skipped.
- Pay attention to the == sign. This is a comparison operator. We use the double equal sign (==) to compare the equality of two operands.
- Also, notice the colon (:) after the if condition. The colon is required in Python to mark the end of the condition and tell Python that it should evaluate the condition and decide whether to execute the code inside the if block if the condition is `True`.
- Finally, notice the indentation of the print statement right under the if condition (as shown in the figure below). In Python, you need to indent any lines of code that are part of a conditional block. You can indent the code using the Tab key or spaces. This tells Python that the indented lines of code are part of the if statement, and they will only run if the condition is `True`.

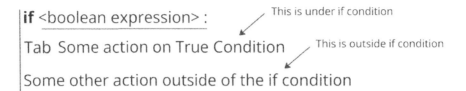

- If the condition in a Python if statement is not `True`, the program doesn't stop. Instead, it moves on and continues to execute the rest of the code. For example, if the condition is `False`, Python will simply skip over the code inside the if block and proceed to execute the code that follows it. This is the reason, the code printed `Amelia's shirt is not red`.

Else Condition

You just saw how we can control the flow of our code using the if condition. It allows us to send the code down a different path if the condition is true, but if the condition is false, it simply continues on the original path. But what if we don't want the code to just continue and instead want to explicitly define what happens when the condition is false? Wouldn't it be better to specify exactly what should happen in the case of a false condition rather than leaving it to Python to skip over the code?

There are also situations in your code where you may want to do something when the condition is false before the code continues along its original path. To handle these situations, Python provides another condition called the else statement. The else block allows you to define what happens when the if condition is false. It gives you control over the flow of the program, making the behavior more predictable and clearer as shown below.

```
color_of_shirt = 'green'
```

```
if color_of_shirt == 'red':
    print("Amelia's shirt is red.")
else:
    print("Amelia's shirt is", color_of_shirt)
print('''
The code came out of the if-else statement
and resumed on the original execution.
''')
----Output----------------
Amelia's shirt is green

The code came out of the if-else statement
and resumed on the original execution.
```

Multiple Conditions (if-elif)

Now, let's take it a step further. What if you want your program to execute multiple statements, not just one? For example, what if you want to say, "If it snows tomorrow, then we will go snow sledding with friends in our backyard; else, if it rains, we will play with paper boats; otherwise, we will go to school."

In such cases, you can use `elif` (short for "else if") to check multiple conditions in sequence as shown below. This allows you to create more complex decision-making paths in your code.

```
if <boolean expression 1> :
Tab  Some action 1
elif <boolean expression 2> :
Tab  Some action 2
else:
Tab  Some action 3

Some other action outside of the if condition
```

Let's take a look at the code example below, which demonstrates how to implement this.

```
color_of_shirt = 'green'
if color_of_shirt == 'red':
    print("Amelia's shirt is red.")
elif color_of_shirt == 'green':
    print("Amelia's shirt is green.")
else:
    print("Amelia's shirt is", color_of_shirt)
----Output----------------
Amelia's shirt is green.
```

Logical Operators in Conditionals

Logical operators (and, or, not) are very commonly used in conditionals to combine multiple conditions.

```
age = 18
score = 50
if age >= 18 and score > 30:
    print("You are eligible for a reward.")
else:
```

```
    print("You are not eligible for a reward.")
----Output----------------
You are eligible for a reward.
```

Nested Conditionals

There might be situations where you need to use a conditional inside another conditional. For example, let's say you want to go snow sledding with friends in your backyard if it's going to snow tomorrow, but only if the temperature is more than 32°F. How can you do that? This is where nested conditionals come in. Let's take a look at the example shown below to understand how they work.

```
will_it_snow = True
temperature = 24
if will_it_snow == True:
    if temperature > 32:
        print("I'll go snow sledding.")
    else:
        print("I'll watch it snow.")
else:
    print("I'll play with paper boats.")
----Output----------------
I'll watch it snow.
```

Let's represent the above flow using a flowchart to make it clearer.

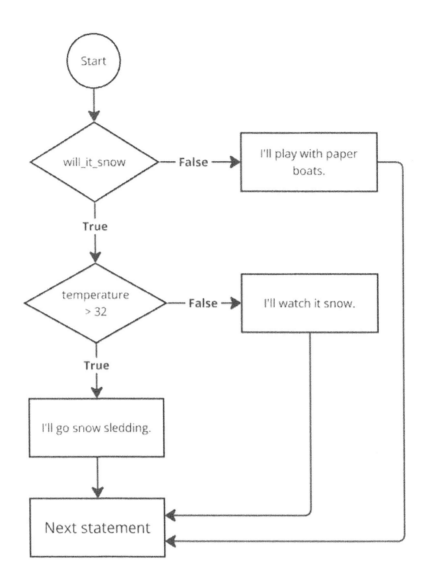

Truthiness (Truthy Vs Falsy)

Sometimes, you may notice that we use a variable directly in an if condition without explicitly comparing it using a comparison operator. For example, if `will_snow = True`, we don't need to write:

```
if will_snow == True:
```

instead, we can simply write,

```
if will_snow:
```

and the code will evaluate `will_snow` and act accordingly.

In the example above, `will_snow` is a variable that we defined, but in Python, there are certain values that can be used directly in conditionals without needing to explicitly compare them to True or False. If a value evaluates to True, we call it **truthy**, and if it evaluates to False, we call it **falsy**.

Following are some common falsy values in Python. Their usage is shown in the code below:
- None
- False (boolean)
- 0 (integer)
- 0.0 (float)
- "" (empty string)

Any other value is considered truthy. There are a few more falsy values in Python that we haven't covered yet, but we'll learn about them in later chapters.

```
string_value = ""
if not string_value:
    print("The string is empty (falsy).")
integer_value = 0
if not integer_value:
    print("Zero is considered falsy.")
my_var = None
if not my_var:
```

```
    print("None is considered falsy.")
----Output----------------
The string is empty (falsy).
Zero is considered falsy.
None is considered falsy.
```

Activity: Let's do an activity! We'll write a Python program to create a Rock, Paper, Scissors game[1]. The game will ask two players to simultaneously choose one of the three hand gestures: "rock," "paper," or "scissors." The winner will be determined based on the rules shown in the diagram below.

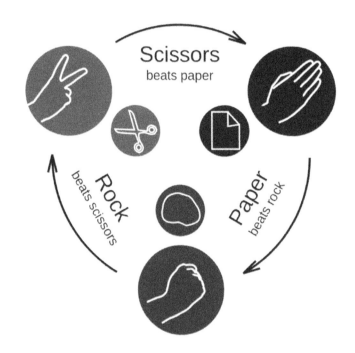

---

[1] https://en.wikipedia.org/wiki/Rock_paper_scissors

The best way to approach this is by creating a flowchart to visualize and understand the conditionals involved.

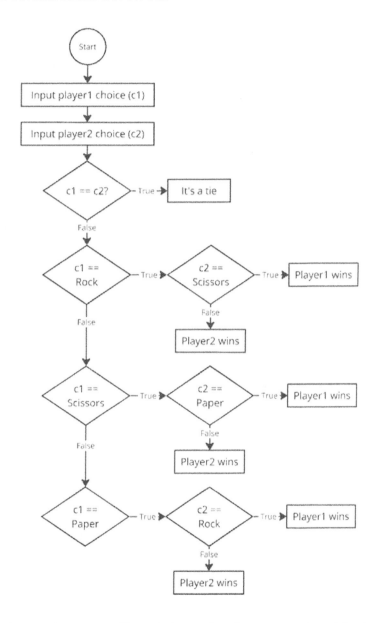

First try writing the code yourself and then check your code with the following approach:

```python
# Game begins
print("Rock, Paper, Scissors Game!")

# Getting input from both players
player1_choice = input("Player 1, enter your choice (rock, paper,
scissors): ").lower()
player2_choice = input("Player 2, enter your choice (rock, paper,
scissors): ").lower()

# Determine the result based on the rules

# Check if it's a tie
if player1_choice == player2_choice:
    print("It's a tie!")
else:
    # Check for Player 1's win conditions
    if player1_choice == "rock":
        if player2_choice == "scissors":
            print("Player 1 wins!")
        else:
            print("Player 2 wins!")
    elif player1_choice == "scissors":
        if player2_choice == "paper":
            print("Player 1 wins!")
        else:
            print("Player 2 wins!")
    elif player1_choice == "paper":
        if player2_choice == "rock":
            print("Player 1 wins!")
        else:
            print("Player 2 wins!")
```

Review Questions

Read the questions carefully and select the correct options to assess how much you have learned so far. Note that more than one answer may be correct.

1. Which of the following are considered falsy values in Python?
    a) `0`
    b) `""` (empty string)
    c) `[]` (empty list)
    d) `"Hello"` (non-empty string)
    e) `False`

2. Which of the following are true about Python's elif statement?
    a) elif allows you to check multiple conditions in sequence.
    b) The elif block will execute only if all preceding if and elif conditions are false.
    c) elif can be used after an else block.
    d) An elif condition can be used without an initial if block.

3. What will happen if you remove the else block from the following code?

```
x = 5
if x < 10:
    print("Less than 10")
else:
    print("Greater than or equal to 10")
```

    a) The code will raise an `IndentationError`.
    b) The code will still work, but it will not print anything when `x` is 5.

c) The code will raise a `ValueError`.

d) The program will run normally and will print `Less than 10`.

4. Which of the following can be used as conditions in Python's if statement?
    a) Boolean values (True/False)
    b) Strings
    c) Lists
    d) Integer values
    e) None of the above

5. Which of the following Python code snippets will evaluate to True?
    a) `if not ""`
    b) `if not 1`
    c) `if 0`
    d) `if 2`
    e) `if "Hello"`

Answer Keys:
1. a, b, c. e; 2. a, b; 3. d; 4. a, b, c, d; 5. a, d, e

12

Loops

Repeat and Automate

Have you ever done something over and over again, like jumping rope or clapping to a beat? In Python, you can make your code run in a similar repetitive way. You might wonder, "Why would I want to repeat things in my code?" Well, running code repeatedly can be super helpful when you want to make something happen multiple times - like printing numbers, drawing shapes, or even repeatedly checking the weather condition, without having to write the same instructions over and over again.

Let's say you're helping a kid draw a square with a pencil. You might say:

"Move the pencil forward 1 inch, then turn 90 degrees."
"Move the pencil forward 1 inch, then turn 90 degrees."
"Move the pencil forward 1 inch, then turn 90 degrees."
"Move the pencil forward 1 inch, then turn 90 degrees."

Did you find this process boring and tedious? You needed to give the same instruction four times for the kid to draw the square. Now, if you think that was tedious then what would you say if I ask you to print `"Hello, World!"` 1000 times?

Loops in Python let you repeat a set of instructions without having to write them out over and over. They make your code more efficient and easier to manage.

In this chapter, we'll explore how loops can make your code work smarter by automating repetitive tasks for you.

Let's look at an example. If you want to print a multiplication table for 2 up to 2 times 5, you'd typically code each line manually, like below:

```
print("2 x 1 =", 2 * 1)
print("2 x 2 =", 2 * 2)
print("2 x 3 =", 2 * 3)
print("2 x 4 =", 2 * 4)
print("2 x 5 =", 2 * 5)
----Output-----------------
2 x 1 = 2
2 x 2 = 4
2 x 3 = 6
2 x 4 = 8
2 x 5 = 10
```

But notice how much repetition there is. What if I told you that you can do it in just one line instead of five? That's where loops come in. They allow you to repeat code automatically.

We can use a for loop to automate this repetition and make our code more efficient, as shown below:

```
for n in range(1, 6):
    print(f"2 x {n} =", 2 * n)
----Output-----------------
2 x 1 = 2
2 x 2 = 4
2 x 3 = 6
2 x 4 = 8
2 x 5 = 10
```

Let's take a closer look at the range function used in the above code to understand how it works. After that, we'll explore loops and how they operate.

range() Function

Before we dive into explaining loops and their different types, let's first understand a built-in function `range()`. This will make it easier to understand the for-loop.

The `range()` function helps generate a sequence of numbers within a specified range. For example, if you ask the function for a sequence of numbers starting from 0, incrementing by 1, and going up to (but not including) 5, it will give you the sequence 0, 1, 2, 3, 4. This is really useful when running loops, and we will see how it works in the next section.

The `range()` function takes three arguments:

- start – The starting value of the sequence. This is optional, and if you don't provide it, it defaults to 0.
- stop – The value where the sequence stops but does not include. This is required.
- step – The increment (or decrement) between numbers in the sequence. This is optional, and if not provided, it defaults to 1. If you provide a negative number, the sequence will count backward.

So, the syntax for defining the `range()` function is:

`range(start, stop, step)`

Examples:
- `range(5)` will generate the sequence: 0, 1, 2, 3, 4. Since start and step were not provided, they default to 0 and 1, respectively.

```
print(list(range(5)))
----Output----------------
[0, 1, 2, 3, 4]
```

Note that in the code above, the `range()` function generates the sequence of numbers, and the `list()` function converts that sequence into a list.

- `range(1, 5)` will generate: 1, 2, 3, 4. Here, start is 1, and step defaults to 1.

```
print(list(range(1,5)))
----Output----------------
[1, 2, 3, 4]
```

- `range(1, 5, 2)` will generate: 1, 3. In this case, the step is 2, so it increments by 2, but stops before reaching 5, since the sequence goes up to, but does not include the stop value.

```
print(list(range(1,5,2)))
----Output----------------
[1, 3]
```

Now, let's explore the two main types of loops: for loops and while loops.

For Loop

This type of loop is used when you know exactly how many times you need to repeat an action. For example, if you have a list and want to go through each element, a for loop is perfect for this task.

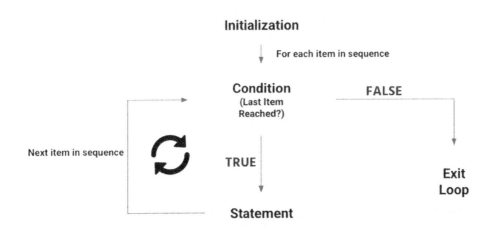

To write a for loop, you specify the starting point, the operation to perform, and the stopping condition. For instance, you could say: 'Start with number 1, multiply it by 2, print the result, increase the number by 1, and repeat until the number reaches 5.' This is exactly what we did in the code in the previous section.

Now let's take another example where you have a list of your friends:

```
friends = ["Amelia", "Henry", "Miles", "Jack"]
```

You want to print greeting for each of these friends like `"Hi, Amelia!"`, `"Hi, Henry!"`, and so on.

A for-loop for this can be written like:

```
friends = ["Amelia", "Henry", "Miles", "Jack"]
for friend in friends:
    print(f"Hi, {friend}!")
----Output----------------
Hi, Amelia!
Hi, Henry!
Hi, Miles!
Hi, Jack!
```

The above code is essentially saying: for each name in the friends list, store the current name in the variable friend and then print the value of friend after formatting it with additional text.

Earlier in the Conditionals chapter, we discussed the importance of indentation in Python. The same rule applies to loops such as for and while. Just like with if statements, any lines of code that are part of a loop must be indented. This indentation tells Python that those lines are inside the loop block and will only be executed as part of the loop's iterations. You can use the Tab key or spaces to indent the code. For example, in a for or while loop, all statements that should run repeatedly must be indented under the loop header. If the indentation is missing or incorrect, Python will throw an `IndentationError`. It's important to remember that proper indentation ensures the loop works as expected and avoids logical errors in your code.

If you ever need to loop through numbers, you can use the `range()` function we learned above:

```
for num in range(1,6):
```

```
    print(f"The current number in the loop is {num}.")
----Output----------------
The current number in the loop is 1.
The current number in the loop is 2.
The current number in the loop is 3.
The current number in the loop is 4.
The current number in the loop is 5.
```

 **Activity**: Let's do an activity! You have a list of three friends – "`Amelia`", "`Henry`", and "`Miles`". You have two chairs, numbered one and two. Using for-loop, print all the possible combinations in which you can seat a pair of your friends in these chairs.

friends = ["Amelia", "Henry", "Miles"]

You can first try solving it yourself, and then compare your approach with the following solution:

```
friends = ["Amelia", "Henry", "Miles"]
for i in friends:
    for j in friends:
        if j != i:
            print(f"Chair 1: {i}, Chair 2: {j}")
----Output----------------
Chair 1: Amelia, Chair 2: Henry
Chair 1: Amelia, Chair 2: Miles
Chair 1: Henry, Chair 2: Amelia
```

```
Chair 1: Henry, Chair 2: Miles
Chair 1: Miles, Chair 2: Amelia
Chair 1: Miles, Chair 2: Henry
```

We have used a nested for loop here. The first for loop goes through each friend in the list, and for each friend, the second for loop also goes through the list. In the second for loop, we check if the two friends picked are not the same (using the condition if `j != i:`) to avoid assigning the same friend to both chairs. This ensures that each combination consists of two different friends.

While Loop

A while loop is used when you want to keep repeating a set of actions until a specific condition is met. The loop will continue as long as the condition is true. As soon as the condition becomes false, the loop stops, and the code execution moves on to the next statement of the program.

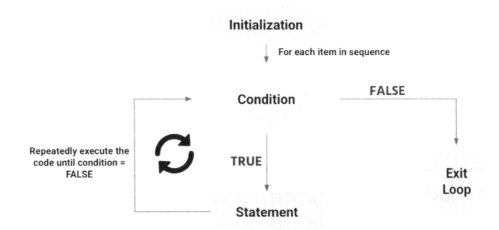

Remember, if the condition of the loop never becomes false, it will result in an infinite loop, causing the program to keep running without ever exiting the loop.

Now, let's look at an example to demonstrate how a while loop works. Suppose you need to count down for a rocket launch, starting at 10. When you reach 0, the rocket will blast off. Try running the code below in your Mu Editor.

```
import time
count = 10
while count > 0:
    print(count)
    time.sleep(1)
    count -= 1
print("Rocket blasting off")
----Output----------------
10
9
8
7
6
5
4
3
2
1
Rocket blasting off
```

In the above code, we can ignore the lines `import time` and `time.sleep(1)`. These lines are used to add a 1-second delay before printing each countdown value. This allows us to pause for a brief moment between each count, making the countdown appear more realistic, like a real rocket launch countdown.

We will learn more about the import statement and how to use modules like time later in this book.

In the above code, you are initially assigning a value of 10 to a variable `count`. Then, you tell the while loop to check the condition `count > 0`. If this condition is true, the loop will execute. Initially, the condition will be true because count is 10 and `10 > 0`, so the code enters the loop.

Inside the loop, the counter is printed, and then the value of count is decremented by `1`. The updated value is reassigned to count. The loop repeats, and the condition is checked again. For example, the next time the condition will be `9 > 0`, which is true, so the loop executes again.

This continues until count becomes 0. At this point, the condition `count > 0` evaluates to `False`, and the loop exits. The final statement, `"Rocket blasting off"`, is then printed.

Activity: Can you think of another scenario where a while loop can be effectively used? How about the Rock, Paper, Scissors game we coded earlier in the Conditionals chapter? What if we want to improve our code by ensuring that players only enter one of the valid choices — "rock," "paper," or "scissors"?

This is a great use case for a while loop. We can keep prompting the player for input until they enter a valid choice. Let's see how we can implement this. We will be implementing it only for player1 and then you can implement rest of the code for `player2` applying a similar logic.

```
valid_choices = ["rock", "paper", "scissors"]
```

```
player1_choice = ""
while player1_choice not in valid_choices:
    player1_choice = input("Player 1, enter your choice (rock, paper,
scissors): ").lower()
    if player1_choice not in valid_choices:
        print("Invalid choice! Please choose from 'rock', 'paper', or
'scissors'.")
print(f"Player 1 chose: {player1_choice}")
```

Run this code and observe how it works. First, try entering an incorrect value, and then provide one of the valid choices.

Note that we have used the condition `player1_choice` **not in** `valid_choices`. This checks whether the value assigned to the variable `player1_choice` exists in the list `valid_choices`. If the value is not found in the list, the program enters the while loop and keeps prompting the user for a valid input. Once a valid choice is entered, the loop exits, and the program proceeds to print the valid choice.

Review Questions

Read the questions carefully and select the correct options to assess how much you have learned so far. Note that more than one answer may be correct.

1. What will the following code print?

    ```
    numbers = [1, 2, 3, 4, 5]
    for n in numbers:
        if n % 2 == 0:
            print("Even")
        else:
            break
    ```

    a) Even
    b) Even followed by an error
    c) Even followed by the program halting
    d) Even, followed by a break in the loop

2. What is the result of the following code snippet?

    ```
    for i in range(5, 0, -1):
        print(i)
    ```

    a) 5 4 3 2 1
    b) 5 4 3 2
    c) 4 3 2 1
    d) 1 2 3 4 5

3. What is the output of the following code?

    ```
    count = 0
    while count < 5:
        print(count)
        count += 2
    ```

    a) 0 2 4
    b) 1 3
    c) 0 1 2 3 4
    d) Infinite loop

4. Which of the following scenarios would be a good use case for a while loop?
    a) Iterating over a list of numbers and performing a fixed operation on each one.
    b) Counting down from 10 to 1, printing each number, and stopping when you reach 1.
    c) Calculating the sum of numbers in a predefined list of known size.
    d) Repeating a task until a specific condition, like entering a correct password, is met.

5. What will the following code output?

    ```
    for i in range(0, 5, 2):
        print(i, end=" ")
    ```

    a) 0 1 2 3 4
    b) 0 2 4
    c) 1 3

d) 0 2 4 6

6. What will this while loop do?

```
n = 0
while n < 10:
    print(n)
    if n == 5:
        break
    n += 2
```

a) Print 0, 2, 4, 6

b) Print 0, 2, 4, 6, 8

c) Print 0, 2, 4, 6, and then exit

d) Print an infinite loop of numbers

7. Which of the following are valid uses of the `range()` function in Python?
   a) range(10, 1, -1)
   b) range(1, 10)
   c) range(1, 10, 2)
   d) range(10, 1, 1)

8. Consider the following code. What will happen when it is run?

```
i = 0
while i < 3:
    print(i)
    i += 1
else:
    print("End of loop")
```

a) It will print 0, 1, 2 and then "End of loop".
b) It will print 0, 1, 2 and then exit without printing anything further.
c) It will cause an infinite loop.
d) The program will throw an `IndentationError`.

9. Which of the following is true regarding nested loops?
    a) A for loop inside a while loop can be used to iterate over a fixed sequence of numbers while also checking an ongoing condition.
    b) A while loop inside a for loop is invalid and will throw an error.
    c) Nested loops allow us to access more complex data structures such as multi-dimensional lists.
    d) Nested loops always run faster than non-nested loops.

Answer Keys:
1. c; 2. a; 3. a; 4. b, d; 5. b; 6. b; 7. a, b, c; 8. a; 9. a, c

13

Functions

The Magical, Relentless Helper

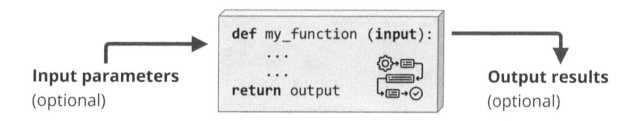

In Chapter 3, we covered the basics of Python functions—what they are, how they work, and why they're important in programming. We also explored a few built-in functions, like `print()`, and in later chapters, we learned about several other built-in functions. In this chapter, we'll learn how to create our own functions to make our programs more reusable and efficient.

Let's first take a look at the following code and see what it does:

```
import random

num = int(input("Enter a number: "))
```

```
if num % 2 == 0:
    print(num, "is even")
else:
    print(num, "is odd")

num = num + random.randint(1, 10)

print("New randomized number is: ", num)

if num % 2 == 0:
    print(num, "is even")
else:
    print(num, "is odd")
----Output----------------
Enter a number: 5
5 is odd
New randomized number is:  12
12 is even
```

Let's break it down to understand what this code does:
1. The program starts by asking the user to enter a number using `input()`. The entered value is then converted into an integer using `int()`.
2. The program checks whether the entered number is even or odd by using the modulus operator (`%`).
    - If `num % 2 == 0`, the number is even.
    - If not, the number is odd.
3. The program then generates a random number between 1 and 10 using `random.randint(1, 10)` and adds it to the original num.
4. The program repeats the step#2 above to checks whether the new randomized number is even or odd and prints the result.

What's wrong with the above code?

Technically, there's nothing wrong with the code, but we can say it's inefficient. The reason is that we're checking whether the number is even or odd twice—once for the original number and again for the new randomized number. While this works fine, it's not the most efficient approach because:

1. The logic for checking if a number is even or odd is duplicated. If you need to check multiple numbers (or perform this check in multiple places in your program), you'd be repeating the same logic each time, which is unnecessary and inefficient.
2. If you need to modify how the even/odd check works (for example, by changing the print message from `print(num, "is odd")` to `print(num, "is not even"))`, you'd have to update it everywhere the logic is repeated. This increases the chance for mistakes and makes your code harder to maintain in the long run.

Now, Let's look at the same code but updated to fix the two issues discussed above:

```python
import random

def check_even_odd(num):
    if num % 2 == 0:
        print(num, "is even")
    else:
        print(num, "is odd")

num = int(input("Enter a number: "))

check_even_odd(num)

num = num + random.randint(1, 10)
```

```
print("New randomized number is: ", num)
check_even_odd(num)
----Output----------------
Enter a number: 5
5 is odd
New randomized number is:  12
12 is even
```

What difference do you notice in this modified code?

In the modified code, instead of repeating the logic to check if a number is even or odd, we've encapsulated it in a function. Whenever we need to check if a given number is even or odd, we simply call the `check_even_odd(num)` function and pass the number as an argument.

You can see that we've called this function twice—once for the original number and once for the randomized number. The benefit is that the `check_even_odd()` function can be called multiple times throughout the program whenever you need to perform this check. This eliminates the need to duplicate the logic each time.

Now, imagine you want to change the print message. For example, if you want to change the message from `print(num, "is odd")` to `print(num, "is not even")`, you only have to update it in the `check_even_odd()` function. You won't have to modify it in every place the logic is repeated. This makes your code easier to maintain and less error prone.

In addition, the main part of the program is now cleaner and more readable, as the repetitive logic is neatly contained within the `check_even_odd()` function.

Function Definition

In the previous section, you saw how we created a reusable function to check if a number is even or odd. Did you notice that the function we created has a specific structure? Take a look at the diagram below to understand the structure of a function definition.

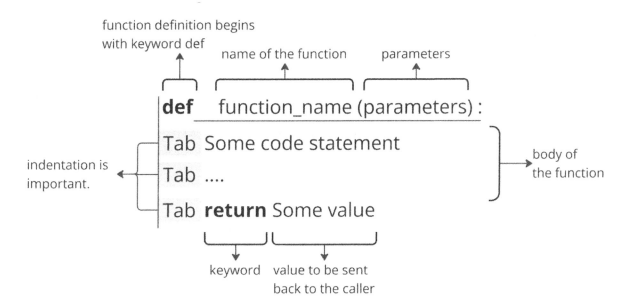

A function definition begins with the **def** keyword. This stands for "define" and tells Python that you are creating a function. After the def keyword, you provide a name for your function. In the example above, the function name is `check_even_odd`.

In the parentheses `()`, you define the parameters (also called arguments when passed from the function call) of the function. These are the values the function will work with when it is called. After the parentheses, a colon (`:`) is used to indicate the start of the function's body. This is mandatory in Python and marks the beginning of the indented block that contains the logic of the function.

The body of the function contains the actual code that is executed when the function is called. It is indented (usually by 4 spaces or a tab) to separate it from the rest of the code. In this case, the body of the function includes an if-else statement that checks whether the number num is even or odd.

A function can also return a value using the `return` keyword. When Python reaches the return statement, it stops executing the function and sends the result back to the caller. If no return statement is provided, the function returns None by default.

In the example above, we do not have a return statement, as the function simply prints the result. However, in another case, if we need to write a function that adds two numbers, the code might look like this:

```
def add_numbers(a, b):
    return a + b
print(add_numbers(2,3))
----Output----------------
5
```

In this example, the `add_numbers()` function returns the sum of a and b when called.

Calling a Function

To call a function, you use the function name followed by parentheses like shown below:

```
add_numbers()
```

Parameters and Arguments

We saw in the previous section that parameters are variables that are defined in the function declaration, and they act as placeholders for values you'll pass when calling the function. Arguments are different. They are the actual values you pass to a function when you call it. In the below code, the function `add_numbers()` is called by passing arguments 3 and 5. These arguments will then be mapped to the function parameters a and b respectively.

```
print(add_numbers(3, 5))
```

Return Statement

If you want your function to send you back the result, you use the `return` keyword. For example, if you want to divide 3 by 2 in a function, return that result, and then multiply the returned value by 5, you can do so as shown below:

```
def divide_numbers(a, b):
    return a / b
result = divide_numbers(3, 2)
result *= 5
print(result)
----Output-----------------
7.5
```

All functions return results. If you do not explicitly return anything then it returns `None` by default.

```
def divide_numbers(a, b):
    print(a / b)
```

```
result = divide_numbers(3, 2)
print(result)
----Output----------------
1.5
None
```

Look at the code above. The function doesn't return anything by default. So, when you try printing the returned result, it prints `None`.

Default Parameter

When defining a function, you can assign default values to some parameters. These default values are used when no argument is passed for that parameter.

```
def multiply(a, b=5):
    return a * b
```

In the example code above, `b` is an optional parameter with a default value of `5`. While calling this function `multiply()`, you do not need to pass the parameter `b`. The code will assign `5` to the variable `b` by default.

There are two ways to call this function now:

1. Passing both arguments

    ```
    result = multiply(2, 3)
    print(result)
    ----Output----------------
    6
    ```

2. Omitting the optional argument

```
result = multiply(2)
print(result)
----Output----------------
10
```

Recursive Functions

Can a function call itself?

Though it may sound unreal, it's true! This is called a **recursive** function, and the use of recursive function is called **recursion**. Let's understand this concept with the help of an example. In this example, we will try to count down from any number until we reach 1, and while doing so, we will print all the numbers. For instance, if we start from 5, we will print 5, 4, 3, 2, and 1.

You could easily accomplish this using a for loop, which you learned in the previous chapter. However, for the sake of understanding the concept, we will solve this using recursion.

```
def count_down(n):
    if n == 0:
        print("STOP!")
    else:
        print(n)
        count_down(n - 1)
count_down(5)
----Output----------------
5
4
3
2
1
```

> STOP!

Let's break down the above code to understand what it does:

The function `count_down` has an `if` condition that tells the function when to stop. In this case, when `n` is 0, it prints `STOP!` and exits. The `else` case is where the function recursively calls count_down with a decreasing number passed as an argument each time it is called.

Here's how it works: First, `count_down(5)` is called. It prints 5 and then calls `count_down(4)`. This prints 4 and then calls `count_down(3)`, and so on, until it reaches `count_down(0)`, where `n == 0` becomes true, and the function exits after printing `STOP!`.

Activity: This was a simple explanation of a recursive function. But what if I ask you to rewrite the same function, but this time, instead of printing the numbers each time, store them in a list and return them all together as a list?

I want you to try writing the code for this based on the hints provided below. Once you're done, you can come back and compare your code with the solution given below:

Since you want to store the counted-down numbers this time, you need a list to store these numbers. But where should you define this list? If you define it inside the function, it will get redefined on each recursive call, causing it to lose all the stored values from the previous recursion. How about passing it as a default parameter with an empty list as the default value?

Below is the code that solves this problem:

```
def count_down(n, numbers=[]):
    if n == 0:
        return numbers
    else:
        numbers.append(n)
        return count_down(n - 1, numbers)

result = count_down(5)

print(result)
----Output----------------
[5, 4, 3, 2, 1]
```

Here you can note that the `numbers=[]` argument is initialized once, and it is shared across all recursive calls. This means that the list is modified across different calls of the recursive function. This is why we accumulate all the numbers in the same list.

Documenting Functions

It's always a good practice to document your code whenever possible. To document a function and describe what it does, we use a special kind of comment called a **docstring**. A docstring is placed immediately under the function definition and is a string enclosed in triple double quotes (""" """).

```
def add_numbers(a, b):
    """
    This function takes two numbers, `a` and `b`, as input and returns
    their sum.

    Parameters:
    a (int or float): The first number to add.
    b (int or float): The second number to add.
```

```
Returns:
int or float: The sum of `a` and `b`.
"""
return a + b
```

Activity: We're going to roll a special set of dice, and each die will have its own number of sides. We'll write a function that takes the number of dice and the number of sides on each die as inputs, and then it will return a list with the results of each die roll.

I want you to try writing the code for this based on the hints provided below. Once you're done, you can come back and compare your code with the solution given below:

Hints:

- A die has multiple sides, and when you roll it, it lands on one of those sides. You can think of each roll as picking a random number between 1 and the number of sides on the die.
- You need to write a function that can "roll" the dice. The function should know how many dice you're rolling and how many sides each die has.
- To simulate rolling a die, you'll need a random number. Python's `random.randint()` function can help you here! It picks a random number between 1 and the number of sides on the die. Remember, you will first need to import the random module at the top of your code using:

    ```
    import random
    ```

- Once you've rolled the dice, you'll want to store the results. How can you keep track of each roll? A list would be a great way to collect the numbers that come up.
- Since you need to roll the dice multiple times (one for each die), a loop can help. Can you use a loop to repeat the rolling process and generate a new number for each die?
- Don't forget: the function must take two inputs: how many dice you want to roll, and how many sides each die has. Make sure your function uses these inputs when rolling the dice.

```
import random

def roll_dice(num_of_dice, num_of_sides):
    """
    Rolls a specified number of dice, each with a specified number of sides.

    Parameters:
    num_of_dice (int): The number of dice to roll.
    num_of_sides (int): The number of sides on each die.

    Returns:
    list: A list containing the results of each die roll.
    """
    rolls = []
    for i in range(num_of_dice):
        roll = random.randint(1, num_of_sides)
        rolls.append(roll)
    return rolls

num_of_dice = int(input("How many dice would you like to roll? "))
num_of_sides = int(input("How many sides should each die have? "))
results = roll_dice(num_of_dice, num_of_sides)
print("The results of the rolls are:", results)
```

```
----Output----------------
How many dice would you like to roll? 2
How many sides should each die have? 6
The results of the rolls are: [2, 3]
```

Tracing Function Call

Now, let's take a closer look at what happens when the Python interpreter executes a function call. We'll use the example from our previous activity, where we rolled dice, to understand this better.

STEP 1.   Python loads the random module into memory, making it available for generating random numbers.

```
➡ import random
  def roll_dice(num_of_dice, num_of_sides):
      rolls = []
      for i in range(num_of_dice):
          roll = random.randint(1, num_of_sides)
          rolls.append(roll)
      return rolls
  num_of_dice = int(input("How many dice would you like to roll? "))
  num_of_sides = int(input("How many sides should each die have? "))
  results = roll_dice(num_of_dice, num_of_sides)
  print("The results of the rolls are:", results)
```

Memory	Output / Console
random module	

Chapter 13

STEP 2.

```
import random
➡ def roll_dice(num_of_dice, num_of_sides):
      rolls = []
      for i in range(num_of_dice):
          roll = random.randint(1, num_of_sides)
          rolls.append(roll)
      return rolls
  num_of_dice = int(input("How many dice would you like to roll? "))
  num_of_sides = int(input("How many sides should each die have? "))
  results = roll_dice(num_of_dice, num_of_sides)
  print("The results of the rolls are:", results)
```

Memory	Output / Console
random module roll_dice(num_of_dice, num_of_sides)	

STEP 3.

```
  import random
  def roll_dice(num_of_dice, num_of_sides):
      rolls = []
      for i in range(num_of_dice):
          roll = random.randint(1, num_of_sides)
          rolls.append(roll)
      return rolls
➡ num_of_dice = int(input("How many dice would you like to roll? "))
  num_of_sides = int(input("How many sides should each die have? "))
```

```
results = roll_dice(num_of_dice, num_of_sides)
print("The results of the rolls are:", results)
```

Memory	Output / Console
random module roll_dice(num_of_dice, num_of_sides) num_of_dice = 2	How many dice would you like to roll? 2

STEP 4.

```
import random
def roll_dice(num_of_dice, num_of_sides):
    rolls = []
    for i in range(num_of_dice):
        roll = random.randint(1, num_of_sides)
        rolls.append(roll)
    return rolls
num_of_dice = int(input("How many dice would you like to roll? "))
➤num_of_sides = int(input("How many sides should each die have? "))
results = roll_dice(num_of_dice, num_of_sides)
print("The results of the rolls are:", results)
```

Memory	Output / Console

Memory	Output / Console
random module

roll_dice(num_of_dice, num_of_sides)

num_of_dice = 2

num_of_sides = 6 | How many dice would you like to roll? 2

How many sides should each die have? 6 |

STEP 5.   In this step, the function `roll_dice(2, 6)` is called. The interpreter gets inside the `roll_dice()` function.

```
import random
def roll_dice(num_of_dice, num_of_sides):
    rolls = []
    for i in range(num_of_dice):
        roll = random.randint(1, num_of_sides)
        rolls.append(roll)
    return rolls
num_of_dice = int(input("How many dice would you like to roll? "))
num_of_sides = int(input("How many sides should each die have? "))
➤results = roll_dice(num_of_dice, num_of_sides)
print("The results of the rolls are:", results)
```

Memory	Output / Console
random module

roll_dice(num_of_dice, num_of_sides) | How many dice would you like to roll? 2 |

num_of_dice = 2 num_of_sides = 6	How many sides should each die have? 6

STEP 6.

```
import random
def roll_dice(num_of_dice, num_of_sides):
    rolls = []
    for i in range(num_of_dice):
        roll = random.randint(1, num_of_sides)
        rolls.append(roll)
    return rolls
num_of_dice = int(input("How many dice would you like to roll? "))
num_of_sides = int(input("How many sides should each die have? "))
results = roll_dice(num_of_dice, num_of_sides)
print("The results of the rolls are:", results)
```

Memory	Output / Console
random module roll_dice(num_of_dice, num_of_sides) num_of_dice = 2 num_of_sides = 6 rolls = []	How many dice would you like to roll? 2 How many sides should each die have? 6

Chapter 13

STEP 7. In this step, the for-loop is initialized, and a list of numbers is generated `[0, 1]` using `range(num_of_dice)` and the first element of the list is assigned to a variable `i`.

```
import random
def roll_dice(num_of_dice, num_of_sides):
    rolls = []
    for i in range(num_of_dice):
        roll = random.randint(1, num_of_sides)
        rolls.append(roll)
    return rolls
num_of_dice = int(input("How many dice would you like to roll? "))
num_of_sides = int(input("How many sides should each die have? "))
results = roll_dice(num_of_dice, num_of_sides)
print("The results of the rolls are:", results)
```

Memory	Output / Console
random module roll_dice(num_of_dice, num_of_sides) num_of_dice = 2 num_of_sides = 6 rolls = [] i = 0	How many dice would you like to roll? 2 How many sides should each die have? 6

STEP 8. `random.randint(1, 6)` is called to generate a random int between 1 and 6. Say it generated 4.

```
import random
def roll_dice(num_of_dice, num_of_sides):
    rolls = []
    for i in range(num_of_dice):
        roll = random.randint(1, num_of_sides)
        rolls.append(roll)
    return rolls
num_of_dice = int(input("How many dice would you like to roll? "))
num_of_sides = int(input("How many sides should each die have? "))
results = roll_dice(num_of_dice, num_of_sides)
print("The results of the rolls are:", results)
```

Memory	Output / Console
random module roll_dice(num_of_dice, num_of_sides) num_of_dice = 2 num_of_sides = 6 rolls = [] i = 0 roll = 4	How many dice would you like to roll? 2 How many sides should each die have? 6

STEP 9. The `rolls.append(4)` method adds 4 to the rolls list, so the list is now [4].

```
import random
def roll_dice(num_of_dice, num_of_sides):
    rolls = []
    for i in range(num_of_dice):
        roll = random.randint(1, num_of_sides)
        rolls.append(roll)
    return rolls
num_of_dice = int(input("How many dice would you like to roll? "))
num_of_sides = int(input("How many sides should each die have? "))
results = roll_dice(num_of_dice, num_of_sides)
print("The results of the rolls are:", results)
```

Memory	Output / Console
random module roll_dice(num_of_dice, num_of_sides) num_of_dice = 2 num_of_sides = 6 **rolls = []** i = 0 roll = 4	How many dice would you like to roll? 2 How many sides should each die have? 6

| rolls = [4] | |

STEP 10. The execution has reached the end of the for-loop so it will check if there are more elements to process in the list. Since only the first iteration (i = 0) has been completed, the loop continues with the next iteration (i = 1).

```
import random
def roll_dice(num_of_dice, num_of_sides):
    rolls = []
    for i in range(num_of_dice):
        roll = random.randint(1, num_of_sides)
        rolls.append(roll)
    return rolls
num_of_dice = int(input("How many dice would you like to roll? "))
num_of_sides = int(input("How many sides should each die have? "))
results = roll_dice(num_of_dice, num_of_sides)
print("The results of the rolls are:", results)
```

Memory	Output / Console
random module roll_dice(num_of_dice, num_of_sides) num_of_dice = 2 num_of_sides = 6 **rolls = []**	How many dice would you like to roll? 2 How many sides should each die have? 6

~~i = 0~~ roll = 4 rolls = [4] i = 1	

STEP 11. Now the Steps 8 and 9 will be repeated to update variables `roll` and `rolls` with new values. This time, say the random int generated is 2.

```
import random
def roll_dice(num_of_dice, num_of_sides):
    rolls = []
    for i in range(num_of_dice):
        roll = random.randint(1, num_of_sides)
        rolls.append(roll)
    return rolls
num_of_dice = int(input("How many dice would you like to roll? "))
num_of_sides = int(input("How many sides should each die have? "))
results = roll_dice(num_of_dice, num_of_sides)
print("The results of the rolls are:", results)
```

Memory	Output / Console
random module roll_dice(num_of_dice, num_of_sides)	How many dice would you like to roll? 2

num_of_dice = 2 num_of_sides = 6 ~~rolls = []~~ ~~i = 0~~ ~~roll = 4~~ ~~rolls = [4]~~ i = 1 roll = 2 rolls = [4, 2]	How many sides should each die have? 6

STEP 12. Now, since the loop has iterated through all values from `range(num_of_dice)`, the control exits the for loop by returning the `rolls` variable. This returned value is then assigned to another variable, `results`, as requested in Step 5. Note that the variables `i`, `roll`, and `rolls` created inside the for loop are removed from memory once the loop ends, because Python no longer needs them after the loop has completed.

```
import random
def roll_dice(num_of_dice, num_of_sides):
    rolls = []
    for i in range(num_of_dice):
        roll = random.randint(1, num_of_sides)
        rolls.append(roll)
```

Chapter 13

```
        return rolls
num_of_dice = int(input("How many dice would you like to roll? "))
num_of_sides = int(input("How many sides should each die have? "))
results = roll_dice(num_of_dice, num_of_sides)
print("The results of the rolls are:", results)
```

Memory	Output / Console
random module roll_dice(num_of_dice, num_of_sides) num_of_dice = 2 num_of_sides = 6 ~~i = 1~~ ~~roll = 2~~ ~~rolls = [4, 2]~~ results = [4, 2]	How many dice would you like to roll? 2 How many sides should each die have? 6

STEP 13. Results are now printed on the console.

```
import random
def roll_dice(num_of_dice, num_of_sides):
    rolls = []
    for i in range(num_of_dice):
```

```
        roll = random.randint(1, num_of_sides)
        rolls.append(roll)
    return rolls
num_of_dice = int(input("How many dice would you like to roll? "))
num_of_sides = int(input("How many sides should each die have? "))
results = roll_dice(num_of_dice, num_of_sides)
```
➡ `print("The results of the rolls are:", results)`

Memory	Output / Console
random module roll_dice(num_of_dice, num_of_sides) num_of_dice = 2 num_of_sides = 6 results = [4, 2]	How many dice would you like to roll? 2 How many sides should each die have? 6 The results of the rolls are: [4, 2]

Review Questions

Read the questions carefully and select the correct options to assess how much you have learned so far. Note that more than one answer may be correct.

1. How can you modify a function to return the result instead of just printing it?
    a) Remove the `return` keyword and use `print()` inside the function.
    b) Add a `return` statement at the end of the function to send the result back.
    c) Remove the `def` keyword.
    d) Pass the result directly to the `input()` function.

2. What will be the output of the below code?

    ```
    def greet(name):
        print(f"Hello, {name}!")

    greet("Alice")
    ```

    a) Hello, {name}!
    b) Error: name not defined
    c) Alice
    d) Hello, Alice!

3. Why is recursion useful in programming?
    a) It allows solving problems that can be broken down into smaller, repetitive subproblems

b) It reduces the need for loops in programs

c) It is the only way to solve mathematical problems

d) It is faster than using loops in all cases

4. How do you pass multiple values to a function?
    a) By passing them as a list or tuple inside the function call.
    b) By passing them as arguments inside the parentheses in the function call.
    c) By using the `input()` function inside the function.
    d) By using a single string with comma-separated values.

5. What happens if you call a function before defining it in Python?
    a) It will cause a runtime error.
    b) Python will throw a `NameError` because it doesn't know about the function yet.
    c) It will execute the function normally.
    d) Python will automatically define the function at runtime.

Answer Keys:
1. b; 2. d; 3. a, b; 4. a, b; 5. b

14

Modules

Organize and Reuse

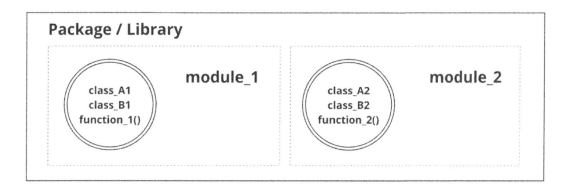

When designing large programs, you'll often face challenges with code organization. As your code grows, it can quickly become difficult to manage. Modular programming helps solve this problem by breaking large programs into smaller, independent files called **modules**. Once these modules are created, they can be easily imported into other programs, just like we did in the previous chapter when we imported the random module by writing import random at the top of our code. This random module contains several functions that are logically grouped together, making the code easier to maintain. We don't need to understand the details of how these functions work. For example, the `randint()` function always returns a random integer, and we don't need to worry about how it achieves that.

I want to talk about an important term: **library**. People sometimes refer to them as **package** as well. While packages are a slightly different concept, for now, you can use these terms interchangeably. You might hear the words "module" and "library" a lot in Python, and while they seem similar, they are actually different. A module is simply one file in Python that has a bunch of related functions, classes, and variables. It's a `.py` file that you can import into your program to use the things it contains. For example, the random module helps you generate random numbers, and the math module gives you different math functions like square roots. On the other hand, a library is a collection of several related modules, all grouped together to help with a specific task. Think of it like a toolbox, where each module is a different tool. A library can have many modules, sometimes even hundreds, all working together for a common purpose. So, while a module is a single tool, a library is a big set of tools.

Types of Modules

There are primarily three types of Python modules:

- **Built-in modules** come pre-installed with Python like random (has functions to generate random numbers), math (has many mathematical functions), etc.

```
import random

random_number = random.randint(1, 10)
```

```
print(random_number)
```

- **Third-party modules** are external modules that needs to be installed separately like *numpy* (used for numerical computing), pandas (for data manipulation and analysis), etc. These modules are part of libraries that are needed to be installed using a command called `pip`. One example of how to install a third-party library, such as requests, is given below.

```
pip install requests
```

 If you are using Mu Editor, you do not need to use `pip` for installing third party modules. Just follow the instructions given at the url https://codewith.mu/en/tutorials/1.2/pypi It gives step-by-step instructions for installing them.

After you install the requests module, try the following code and see what you get:

```
import requests

response = requests.get('https://www.example.com')

print(response.text)
```

- **User-defined modules** are modules that you can create yourself. These are `.py` files that contain functions, classes, and variables, which can be reused in other Python programs. In this chapter, we will explore how to create your own modules and see how you can organize your code into reusable pieces.

Importing Modules

Before we discuss specific types of modules, let's first understand how to import and access modules in our code. You need to import a module before you can access its classes, functions, and variables. There are different ways to import a module. We've already seen one basic method: `import <module_name>`. Now, let's explore two other main methods:

Import with Alias

Sometimes, you might want to import a module but don't like its default name, or you just want to shorten it. You can rename a module when importing it by using the **as** keyword, like this:

```
import math as m
print(m.pi)
```

In this example, we've renamed the built-in math module to `m`, making it easier to reference throughout the code.

Import Specific Functions or Classes

Sometimes, you might not want to import the entire module but only a specific function or item from it. For example, if you only need the pi constant from the math module, you can import just that using the **from** keyword, like this:

```
from math import pi
print(pi)
```

In this case, we've imported only the pi constant from the math module, so you don't have to reference the entire module.

User Defined Modules

Now that we've discussed what modules are and the different types of modules, let's dive into user-defined modules. We'll learn by creating a simple example module and reusing it. Our module will be named `my_math_calculator`, and it will contain two functions: one for adding two numbers and another for multiplying two numbers. In addition, we'll define a constant called ZERO with a value of 0. This allows us to use ZERO in place of writing the number 0, making our code more readable and reusable.

First, create a new file in your Mu Editor and name it `my_math_calculator.py`. In this file, write your constant and the two function definitions as shown below:

```
#------- my_math_calculator.py----
ZERO = 0
def add(a, b):
    return a + b
def multiply(a, b):
    return a * b
```

Next, create another file named `main.py`. This is the code file that will import `my_math_calculator` module and use the functions inside it.

```
#------- main.py--------
import my_math_calculator as m

print(m.add(2,3))
print(m.multiply(2,3))
print(m.ZERO)
```

Now, run your code in `main.py`. You should see the following output:

```
5
6
0
```

Use __name__ and __main__ to Avoid Running Code As a Script

Let's modify the above code to understand how we avoid running certain parts of code when a module is imported but still allow it to run when executed as a script:

```
#------- my_math_calculator.py----
ZERO = 0
def add(a, b):
    return a + b
def multiply(a, b):
    return a * b
if __name__ == "__main__":
    print("This is the 'main' block in my_math_calculator.py.")
```

When `my_math_calculator.py` is imported as a module, the `__name__` variable inside `my_math_calculator.py` will **not** be '`__main__`'. Instead, it will be '`my_math_calculator`' because Python assigns the module's name to `__name__` when the file is imported.

Since the `if __name__ == "__main__"` condition checks if the file is being executed directly (`__name__` equals '`__main__`'), the code inside this block will not run when `my_math_calculator.py` is imported in `main.py`.

Review Questions

Read the questions carefully and select the correct options to assess how much you have learned so far. Note that more than one answer may be correct.

1. Which of the following is true about Python modules and libraries?
    a) A library can contain multiple modules.
    b) A module can contain multiple libraries.
    c) A module is always a collection of third-party packages.
    d) Libraries are collections of Python files (`.py` files).

2. What happens if you try to import a third-party module that is not installed?
    a) Python will automatically install the module for you.
    b) Python will raise an `ImportError`.
    c) Python will create a new module with the same name.
    d) The code will run without issues, but the module's functionality will be unavailable.

3. What will the following code print?

    ```
    import random
    print(random.choice([1, 2, 3, 4, 5]))
    ```

    a) A random number between 1 and 5
    b) A random number between 0 and 5
    c) A random element from the list [1, 2, 3, 4, 5]

d) 1

4. Which of the following is NOT a valid way to import a module in Python?

    a) import random

    b) import math as m

    c) from random import *

    d) import random()

5. Which statement about user-defined modules is true?

    a) You cannot import functions from a user-defined module into another script.

    b) The module's filename must be exactly the same as the function name.

    c) User-defined modules are simply Python files that contain reusable functions and variables.

    d) User-defined modules must always contain a class.

Answer Keys:

1. a; 2. b; 3. c; 4. d; 5.c

15

Object-Oriented Programming

Understanding Classes, Objects and Methods

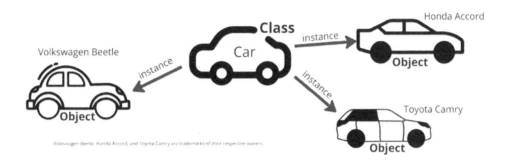

Volkswagen Beetle, Honda Accord, and Toyota Camry are trademarks of their respective owners.

Everything around us can be considered as objects. For example, your parent's car, your computer, your school bag, etc. Have you noticed that many of these objects behave in similar ways? Take your parent's car as an example. I may not know the exact brand or model of your car, but one thing is certain — it moves, has four wheels, brakes, blinks its lights, wipes the rain off the windshield, and can turn left or right. How do I know this? Because all cars share these features. They behave in the same way because they are all cars.

In programming, we use a similar concept. A "car" can be considered a **class**, which defines certain properties (like wheels, brakes, and lights) and behaviors (like moving and turning). Your parent's car is an **object** (also called instance) of the car class. It inherits all the features and behaviors of the car class. But, just like a real car can't fly because it's an object of the "car" class, the same applies in programming: an object can only do what the class allows.

Object-Oriented Programming (OOP) is a coding approach that helps in organizing and structuring your code so that everything is modeled as real-life objects. An object is a fundamental unit of OOP, representing a real-world entity with attributes (characteristics) and methods (actions). Think of objects as LEGO blocks, where each block has its unique shape (attributes) and can perform specific functions (methods). When you create an object in Python, it needs to know what kind of object you want to create. Class is that template that tells Python what object we want to create and how it should behave.

Class and Objects

Let's see how we can instruct Python to create that template (or class) from which objects can be created. To define a class, you have to use the class keyword, provided by Python, then followed by the name of the class and a colon:

```
class Car:
    pass
```

The class name is Car. This is an empty class and has nothing in it. The pass statement in Python is a placeholder. It tells Python that there is no code inside the class for now. pass is used when you need to write a class or function but don't want to add any logic to it yet. Here I just wanted to show you how the class is created. We will implement it later.

Now that we created a Car class, let's see how we can create an object (or instance) of this class and assign it to a variable.

```
class Car:
    pass

my_car = Car()
print(my_car)
----Output-----------------
<__main__.Car object at 0x108ed38e0>
```

As you can see from the above code, we created an object of the class `Car` by using `Car()` and we assigned this object to a variable `my_car`. When I print `my_car`, Python tells us - `my_car` is an object and it is stored in computers memory at address 0x108ed38e0. If you run this same code, you may get a different address in your output. That is normal. You may also create several other objects like below:

```
class Car:
    pass
my_car = Car()
my_neighbors_car = Car()
my_friends_car = Car()

print(my_car, my_neighbors_car, my_friends_car)

if(my_car == my_neighbors_car):
    print("my car is same as my neighbor's car")
else:
    print("my car is different from my neighbor's car")
----Output-----------------
<__main__.Car object at 0x10ae548e0> <__main__.Car object at 0x10ae544c0> <__main__.Car object at 0x10ae54640>
```

```
my car is different from my neighbor's car
```

As you can see from above code example, two objects that are created from the same class are different.

Attributes

Attributes are data (values or properties) that are associated with an object. We can define attributes inside a class in two different ways:

1. Class Attributes: These are common to all instances (objects) created from the class. If you change a class attribute, it affects every object of that class.

    ```
    class Car:
        wheels = 4

    my_car = Car()
    my_neighbors_car = Car()

    print("Number of wheels in my car", my_car.wheels)
    print("Number of wheels in my neighbor's car",
    my_neighbors_car.wheels)
    ----Output----------------
    Number of wheels in my car 4
    Number of wheels in my neighbor's car 4
    ```

    In this example, we access the class attribute wheels by using the object itself followed by a dot and the attribute name, like `my_car.wheels`.
    All `Car` objects have the same value for the wheels attribute, which is 4 in this case. If you change the value of wheels in the class definition, all car objects will reflect that change.

2. **Instance Attributes:** These are unique to each object created from the class. Each object can have different values for these attributes.

```python
class Car:
    wheels = 4

    def __init__(self, brand, color):
        self.brand = brand
        self.color = color
        self.speed = 0

my_car = Car("Ford", "Black")
my_neighbors_car = Car("Honda", "Red")

print(my_car.brand, my_car.color)
print(my_neighbors_car.brand, my_neighbors_car.color)
----Output-----------------
Ford Black
Honda Red
```

In this code, we introduce the `__init__` method, also called the *constructor*. It takes three parameters: self, brand, and color. The `__init__` method is called automatically whenever a new instance of the Car class is created. This method allows us to configure the car's attributes each time it's created. The brand and color attributes are instance attributes. Every car can have a different brand and color. The wheels attribute remains a class attribute, and every car object shares the same value (4). The speed attribute is also an instance attribute with a default value of 0. Each time you create a new `Car` instance; you need to provide values for the brand and color attributes. If you don't, Python will raise an error. The wheels attribute doesn't need to be provided since it's shared by all instances.

Methods

We've seen how a class can have both class attributes and instance attributes. In addition to attributes, a class can also have methods. Methods are essentially functions, but they are defined within a class and are associated with that class. Because of this, we refer to them as methods instead of regular functions.

Look at the following code to see how we add a method to a class.

```
class Car:
    wheels = 4

    def __init__(self, brand, color, speed):
        self.brand = brand
        self.color = color
        self.speed = 0

    def drive(self, speed_increment):
        self.speed += speed_increment
        print(f"The {self.color} {self.brand} car is now going {self.speed} km/h!")
    def stop(self):
        self.speed = 0
        print(f"The {self.color} {self.brand} car has stopped.")


my_car = Car("Ford", "Black")
my_neighbors_car = Car("Honda", "Red")

my_car.drive(10)
my_car.drive(20)
my_car.stop()
----Output----------------
The Black Ford car is now going 10 km/h!
The Black Ford car is now going 30 km/h!
The Black Ford car has stopped.
```

We defined two methods in the Car class — `drive()` and `stop()`. Just like the `__init__` method, these methods also take self as the first argument. The `self` keyword refers to the current object that the method is being called on.

For example, when you call `my_car.drive(10)`, self refers to `my_car` object, and it will increment the speed attribute of `my_car` by 10. Since the initial speed is 0, the new speed will be `0 + 10 = 10`.

If you call `drive(20)` again, it will add 20 to the current speed (which is now 10), making the new speed `10 + 20 = 30`.

In other words, self makes sure that the method operates on the specific instance of the object (like `my_car`), and each object can have its own values for attributes like speed.

Try calling these methods on `my_neighbors_car` object and observe how they behave. You'll see that the methods affect only `my_neighbors_car` object, and not `my_car`, since each object has its own attributes (like speed).

Encapsulation

Encapsulation is a key concept in Object-Oriented Programming (OOP). Let's understand it with the help of our `Car` class example from the previous section. The `Car` class has methods like `drive()` and `stop()`. With encapsulation, you don't need to worry about how these methods work internally, as long as they allow you to drive and stop the car. It's like putting everything related to the car inside a sealed box, protecting it from outside interference.

To interact with the car, you should only use the provided methods (like `drive()` and `stop()`) and not modify the car's internal attributes directly.

But is our `Car` class properly encapsulated? The answer is no. Here's why: If you look at the code, you can still change the car's speed directly by updating the instance attribute, bypassing the `drive()` method.

For example:

```
my_car = Car("Ford", "Black")
my_car.speed = 20

my_car.drive(10)
my_car.drive(20)
my_car.stop()
```

In this case, the speed can be modified directly, which is a problem. This means the class isn't fully encapsulated or protected from unwanted changes. The correct way to update the car's speed should be through the `drive()` method, ensuring that the class controls how its attributes are modified.

To make the `Car` class properly encapsulated, we need to restrict direct access to the internal attributes (like speed) and ensure that they can only be modified through methods (like `drive()` and `stop()`). We can achieve this by making the speed attribute private, which means it can't be accessed directly from outside the class and then providing controlled access to it through methods (like `drive()` and `stop()`).

In Python, we can make an attribute or method **private** by prefixing its name with two underscores (__). Now, let's change the speed attribute to make it private:

```python
class Car:
    wheels = 4

    def __init__(self, brand, color):
        self.brand = brand
        self.color = color
        self._speed = 0

    def drive(self, speed_increment):
        self._speed += speed_increment
        print(f"The {self.color} {self.brand} car is now going {self._speed} km/h!")
    def stop(self):
        self._speed = 0
        print(f"The {self.color} {self.brand} car has stopped.")

my_neighbors_car = Car("Honda", "Red")

my_car = Car("Ford", "Black")
my_car.speed = 20

my_car.drive(10)
my_car.drive(20)
my_car.stop()
----Output----------------
The Black Ford car is now going 10 km/h!
The Black Ford car is now going 30 km/h!
The Black Ford car has stopped.
```

Note that the line of code `my_car.speed = 20`, where we try to directly set the `speed` to 20, will not work anymore. This is because the speed attribute has been encapsulated and is now protected from outside modification. The speed attribute is private (`_speed`), so it cannot be changed directly from outside the class.

The only way to modify the speed is by using the `drive()` method, which provides controlled access to the speed attribute. This ensures that the car's speed can only be adjusted in a valid way, keeping the class protected and maintaining the integrity of its state.

Inheritance

Inheritance is another key concept in Object-Oriented Programming (OOP). Let's understand it using the `Car` class example from the previous section.

Let's imagine we have a regular `Car`, and then we have a `SportsCar` that has all the features of a regular car but with some special features, like `turbo_speed`. In this case, the `SportsCar` class can inherit the attributes and methods of the regular `Car` class, while adding its own unique features.

In Python, inheritance allows a new class (the child class) to reuse code from an existing class (the parent class). The child class inherits all the methods and attributes from the parent class, and you can also add new methods or override existing ones to make the child class special.

Let's understand inheritance with the help of the following example code:

```python
class Car:
    wheels = 4

    def __init__(self, brand, color):
        self.brand = brand
        self.color = color
        self._speed = 0

    def drive(self, speed_increment):
        self._speed += speed_increment
```

```
            print(f"The {self.color} {self.brand} car is now going
{self._speed} km/h!")
    def stop(self):
        self._speed = 0
        print(f"The {self.color} {self.brand} car has stopped.")


class SportsCar(Car):
    def __init__(self, brand, color, turbo_speed):
        super().__init__(brand, color)
        self.turbo_speed = turbo_speed

    def drive_with_turbo(self):
        self._speed += self.turbo_speed
        print(f"The {self.color} {self.brand} sports car is now going
{self._speed} km/h with turbo!")

my_car = Car("Ford", "Black")
my_car.drive(20)
my_car.stop()

my_sports_car = SportsCar("Ferrari", "Red", 50)
my_sports_car.drive(20)
my_sports_car.drive_with_turbo()
my_sports_car.stop()
----Output----------------
The Black Ford car is now going 20 km/h!
The Black Ford car has stopped.
The Red Ferrari car is now going 20 km/h!
The Red Ferrari sports car is now going 70 km/h with turbo!
The Red Ferrari car has stopped.
```

What differences do you notice in the code above? The `super()` function in the `SportsCar` class allows the child class to call methods from the parent class (like `__init__()`), enabling it to reuse common code for initializing attributes such as brand and color.

When we call `my_car.drive(20)`, it increases the speed by 20 km/h and prints the message as expected. The `SportsCar` starts by using the `drive()` method from the `Car` class, so it behaves like a regular car at first. However, it also has its own method, `drive_with_turbo()`, which allows it to go super-fast by adding extra turbo speed.

The `SportsCar` doesn't need to rewrite the `drive()` or `stop()` methods because it inherits them from the `Car` class. This makes the code cleaner, avoids duplication, and allows us to extend or customize functionality only when necessary.

Review Questions

Read the questions carefully and select the correct options to assess how much you have learned so far. Note that more than one answer may be correct.

1. What does the `pass` keyword do inside a class in Python?
    a) It initializes class attributes.
    b) It indicates the class has no code yet.
    c) It automatically creates objects.
    d) It is used to exit a method.

2. Which of the following statements is true about class attributes in Python?
    a) Class attributes are shared by all instances of a class.
    b) Modifying a class attribute will affect only the current instance.
    c) Class attributes cannot be modified after being defined.
    d) Class attributes are unique to each instance.

3. What is the main advantage of using inheritance in object-oriented programming?
    a) It allows a class to borrow methods and attributes from another class, reducing code duplication.
    b) It makes the child class independent of the parent class.
    c) It lets the parent class call methods from the child class.
    d) It allows the child class to modify the constructor of the parent class.

4. How does encapsulation help in object-oriented programming?
    a) It hides the internal details of an object.

b) It ensures that objects can have their own unique methods.
   c) It makes it possible to modify an object's internal attributes directly.
   d) It provides a way to access internal attributes using getters and setters.
5. What happens if two objects of the same class are compared using == in Python?
   a) They will always be considered equal.
   b) It checks whether they refer to the same memory address.
   c) It checks if they have the same attributes.
   d) It checks if they are instances of the same class.

Answer Keys:
1. b; 2. a; 3. a; 4. a, d; 5. b

16

Project – Library Management System

A Real-Life Sample Project

In this chapter, you will implement a real-life project using Python to develop a Library Management System. First, go through the project requirements carefully to understand the task, and then we will think about how to design and implement it.

Requirements

There are two types of users:

1. Librarian – Manages books and users.
2. Normal User – Searches and borrows books.

The system ensures role-based access, meaning librarians and users see different menus. Books can only be issued if available, and users cannot log in unless a librarian adds them first.

The system starts with a default librarian with username admin. There is no password requirement as we want to keep this simple and for learning purpose only.

When the librarian logs in by entering their name, they see the following menu:

1. Add Book – Add new book with title, author, and ISBN.
2. Add Member – Register a new user or another librarian.
3. Remove Member – Delete an existing user.
4. Search Books – Search by title, author, or ISBN (partial match).
5. List Members – View all registered users.
6. Issue Book – Assign a book to a user if available.
7. Return Book – Mark a book as returned.
8. Logout – Logs out of the system so other user can log in

When a normal user logs in, they see the following menu:

1. Search Books – Search by title, author, or ISBN (partial match).
2. Issue Book – Request a book if it's available.

3. Return Book – Return a previously issued book.
4. Logout – Logs out of the system so other user can log in

Other considerations:

- Users cannot login unless added by a librarian.
- Books must be returned before issuing to another user.
- If a book is already issued, a message should inform the user.

Login Process:

- Users enter their username (case-insensitive).
- If found, they can access their respective menus.
- If not found, they must ask a librarian to add them.

Book Search:

- Users can search by title, author, or ISBN.
- Partial match works (e.g., searching "rowl" finds "J.K. Rowling").
- Case-insensitive (e.g., "harry" finds "Harry Potter").

Book Issuing Rules:

- A book can't be issued if it's already borrowed.
- A message informs users if the book is unavailable.
- The system tracks who have borrowed each book.

Design Thinking

Before we start writing our code, let us think who will use it and what they need that we can design and solve.

Role	Needs	Challenges
Librarian	Manage books (add, remove, list, search). Manage users (add, remove). Issue & return books.	Ensuring users are added before login. Keeping track of which books are issued. Preventing duplicate user or book entries.
Normal Users	Find books quickly using title, author, or ISBN. Borrow books (if available). Return books easily.	Search must be simple with partial match support. Ensuring books are returned before issuing again. Making login case-insensitive for a smooth experience.

Putting the needs and challenges for each role in this grid format makes it easy to get a wholistic idea of the design.

Now, let's brainstorm possible solutions and approach so that we can implement each feature efficiently.

Key Problem / Challenge	Idea / Possible Solution
Users must be authenticated based on role.	Use a dictionary to store users in {username: role} format. Ensure usernames are stored in lowercase for case-insensitive login.
Books must be easy to search.	Store books in a list of objects (Book class). Implement search by title, author, and ISBN using substring matching.
Issued books should not be available for others.	Add an `issued_to` field in the Book class. Before issuing, check if `issued_to` is None`.  When returning, reset `issued_to` to None.
Role-Based Menus.	Store users in a list of objects (Member class). After login, show different options for librarians vs. users.

Test Scenarios:

Here, we will be planning how we will test our application once it is built to make sure it's built as per the specification and meets all requirements.

- Test Case 1 – Login System
    - Input: Username = "admin"
    - Expected Output: Librarian menu appears.
- Test Case 2 – Search Book (Partial Match)
    - Input: Searching for "olivia"
    - Expected Output: Returns "Olivia Smith".
- Test Case 3 – Issue a Book That's Already Issued
    - Input: Issue "Harry Potter" to User1, then to User2
    - Expected Output: "This book is already issued."
- Test Case 4 – Case-Insensitive Login
    - Input: Username = "Amelia" or "amelia"
    - Expected Output: Successful login in both cases.
- Test Case 5 – Trying to Log in Before Being Added
    - Input: User tries to log in without being added
    - Expected Output: "User not found. Please contact a librarian."

System Architecture and Design

Now, we have a fair understanding of the requirements, challenges and possible solutions. Now, let's start with building the solution.

Below are the functional components we will be building:

1. User Management – Add, remove, and authenticate users.

2. Book Management – Add, search, issue, and return books.
3. Role-Based Access Control – Different menus for Librarians and Users.
4. Data Storage – Maintain book and member details in-memory (can extend to files).

Book Class:

```
class Book:
    def __init__(self, title, author, isbn):
        self.title = title.lower()    # Case-insensitive
        self.author = author.lower()
        self.isbn = isbn
        self.issued_to = None
```

As you can see above, we have defined attributes to store book details (title, author, ISBN) and for tracking book availability (`issued_to`). We have converted the title and author to lowercase so we can easily do case insensitive search on them.

Member Class:

```
class Member:
    def __init__(self, name, role):
        self.name = name.lower()
        self.role = role   # "Librarian" or "User"
        self.borrowed_books = []
```

We have added attributes for storing user details (name and role) and for tracking borrowed books (`borrowed_books`) in a list.

Library Class:

```
class Library:
    def __init__(self):
        self.books = []  # List of books
        self.members = {"admin": Member("admin", "Librarian")}  #
Default librarian
```

here, we are attributes to store books in a list and users in a dictionary {username: Member}

Now, let's look at the key features and behaviors.

User Login:

- Users log in using their name (case-insensitive).
- If a user doesn't exist, they must be added by a librarian first.
- Librarians get an extended menu with user management options.

Book Management:

- Add Book – Librarians can add books.
- Search Book – Search by title, author, or ISBN (supports partial match).
- Issue Book – Only if not already issued.
- Return Book – Clears the `issued_to` field.

Role-Based Menus:

Menu Items	Librarian Menu	Normal User Menu
Add a Book	✓	
Add a Member	✓	
Remove a Member	✓	

Search Books	✓	✓
List Members	✓	
Issue Book	✓	✓
Return Book	✓	✓
Log out	✓	✓

Error Handling & Edge Cases:
- User not found? → Show error & prompt librarian to add them.
- Book already issued? → Display availability message.
- Partial search not working? → Ensure case-insensitive substring matching.
- Trying to return a book not borrowed? → Show a warning.

Below is the completed code for this application. Try to understand how it is built and identify any missing features. It may have some bugs or defects that you need to find and fix to make it a robust system.

```python
class Book:
    def __init__(self, title, author, isbn):
        self.title = title.lower()  # Store in lowercase for case-insensitive search
        self.author = author.lower()
        self.isbn = isbn
        self.issued_to = None  # Stores the member who borrowed the book

    def __str__(self):
        status = "Issued to " + self.issued_to if self.issued_to else "Available"
        return f"{self.title.title()} by {self.author.title()} (ISBN: {self.isbn}) - {status}"

class Member:
```

```python
    def __init__(self, name, role):
        self.name = name.lower()  # Store in lowercase for case-insensitive lookup
        self.role = role  # 'Librarian' or 'User'
        self.borrowed_books = []

    def __str__(self):
        return f"{self.name.title()} ({self.role})"

class Library:
    def __init__(self):
        self.books = []
        self.members = {"admin": Member("admin", "Librarian")}  # Default librarian

    def add_book(self, title, author, isbn):
        self.books.append(Book(title, author, isbn))
        print(f"Book '{title}' added successfully!")

    def search_book(self, query):
        query = query.lower()  # Convert query to lowercase for case-insensitive search
        found_books = [
            book for book in self.books
            if query in book.title or query in book.author or query in book.isbn
        ]
        if found_books:
            for book in found_books:
                print(book)
        else:
            print("No matching books found.")

    def add_member(self, librarian_name, member_name, role):
        librarian_name, member_name = librarian_name.lower(), member_name.lower()
        if librarian_name not in self.members or self.members[librarian_name].role != "Librarian":
```

```python
            print("Only a librarian can add members!")
            return
        if member_name in self.members:
            print("Member already exists!")
        else:
            self.members[member_name] = Member(member_name, role)
            print(f"Member '{member_name.title()}' added successfully as {role}.")

    def remove_member(self, librarian_name, member_name):
        librarian_name, member_name = librarian_name.lower(), member_name.lower()
        if librarian_name not in self.members or self.members[librarian_name].role != "Librarian":
            print("Only a librarian can remove members!")
            return
        if member_name in self.members:
            del self.members[member_name]
            print(f"Member '{member_name.title()}' removed successfully!")
        else:
            print("Member not found!")

    def list_members(self):
        if self.members:
            for member in self.members.values():
                print(member)
        else:
            print("No members found.")

    def issue_book(self, member_name, book_title):
        member_name, book_title = member_name.lower(), book_title.lower()
        if member_name not in self.members:
            print("Member not found!")
            return

        member = self.members[member_name]
```

```python
            for book in self.books:
                if book.title == book_title:
                    if book.issued_to:
                        print(f"Sorry, this book is already issued to {book.issued_to.title()}.")
                    else:
                        book.issued_to = member.name
                        member.borrowed_books.append(book)
                        print(f"Book '{book.title.title()}' issued to {member.name.title()}.")
                    return
        print("Book not found!")

    def return_book(self, member_name, book_title):
        member_name, book_title = member_name.lower(), book_title.lower()
        if member_name not in self.members:
            print("Member not found!")
            return
        member = self.members[member_name]
        for book in member.borrowed_books:
            if book.title == book_title:
                book.issued_to = None
                member.borrowed_books.remove(book)
                print(f"Book '{book.title.title()}' returned successfully.")
                return
        print("You did not borrow this book!")

def login(library):
    name = input("Enter your username: ").strip().lower()
    if name in library.members:
        return library.members[name]
    else:
        print("User not found! Please ask the librarian to add you.")
        return None

def librarian_menu(library, user):
```

```python
while True:
    print("\nLibrarian Menu:")
    print("1) Add Book")
    print("2) Add Member")
    print("3) Remove Member")
    print("4) Search Book")
    print("5) List Members")
    print("6) Issue Book")
    print("7) Return Book")
    print("8) Logout")
    choice = input("Enter your choice: ")

    if choice == "1":
        title = input("Enter book title: ")
        author = input("Enter author name: ")
        isbn = input("Enter ISBN: ")
        library.add_book(title, author, isbn)

    elif choice == "2":
        name = input("Enter new member name: ")
        role = input("Enter role (Librarian/User): ").capitalize()
        library.add_member(user.name, name, role)

    elif choice == "3":
        name = input("Enter member name to remove: ")
        library.remove_member(user.name, name)

    elif choice == "4":
        query = input("Enter title, author, or ISBN to search: ")
        library.search_book(query)

    elif choice == "5":
        library.list_members()

    elif choice == "6":
        member_name = input("Enter member name: ")
        book_title = input("Enter book title: ")
        library.issue_book(member_name, book_title)
```

```python
        elif choice == "7":
            member_name = input("Enter member name: ")
            book_title = input("Enter book title: ")
            library.return_book(member_name, book_title)

        elif choice == "8":
            print("Logging out...\n")
            break
        else:
            print("Invalid choice! Try again.")

def user_menu(library, user):
    while True:
        print("\nUser Menu:")
        print("1) Search Books")
        print("2) Issue Book")
        print("3) Return Book")
        print("4) Logout")
        choice = input("Enter your choice: ")

        if choice == "1":
            query = input("Enter title, author, or ISBN to search: ")
            library.search_book(query)

        elif choice == "2":
            book_title = input("Enter book title to issue: ")
            library.issue_book(user.name, book_title)

        elif choice == "3":
            book_title = input("Enter book title to return: ")
            library.return_book(user.name, book_title)

        elif choice == "4":
            print("Logging out...\n")
            break
        else:
            print("Invalid choice! Try again.")
```

```python
# Main Program
def main():
    library = Library()
    print("\nWelcome to the Library Management System!")
    print("Login as 'admin' (default librarian) to add users.")

    while True:
        user = login(library)
        if user:
            if user.role == "Librarian":
                librarian_menu(library, user)
            else:
                user_menu(library, user)

if __name__ == "__main__":
    main()
```

If you look at the last line of the code, you'll see that we have used the variable __name__. This is a special variable in Python. When a script runs, Python sets the __name__ variable. If the script is run directly (rather than imported), __name__ is set to "__main__".

Why do we do this? This helps prevent unintended execution when importing a module. If you import a Python file as a module into another file, its functions won't run automatically. The `main()` function is called only when the script is executed directly.

Try playing around with this code and see what improvement you can make.

Index

___

__init__ · 223
__main__ · 216, 247
__name__ · 216, 247

___

.

.py extension · 30

___

A

access each character · 116
Accessing Dictionary Elements · 143
Accessing List Elements · 132
Adding Elements to a List · 133
Adding to Dictionary · 145
Addition (+) · 99
AND · 90
applications · 21
arguments · 54, 187
Arithmetic Operators · 84
as keyword · 214
assign multiple variables · 50
Assignment Operators · 87
Attributes · 222

___

B

Backslash (\) · 113
Boolean · 68

browser-based Python · 26
built-in functions · 55

___

C

capitalize() · 120
ceiling · 102
Chaining Assignments · 88
Changing Case · 120
characteristics of a list · 129
class · 219, 220
Class Attributes · 222
clear() · 146
Clearing a List · 137
code · 16
Code Execution · 75
Coding · 18
Comparison operators · 85
compiled · 20, 34
Concatenation · 114
condition · 90
conditional statements · 153
constructor · 129
Converting Numbers · 97
Creating a Dictionary · 142
Creating a List · 129
Creating a nested list · 131
Creating an empty list · 134
Creating Tuples · 140

___

D

Data Structure · 126

Data Type Conversion · 69
data types · 64
decisions · 153
def keyword · 187
degrees · 38
Design · 238
Design Thinking · 235
dict · 143
Dictionaries · 125
Dictionary · 142
Dictionary Unpacking · 148
Division (/) · 99
docstring · 193
drawing tool · 36

E

Else Condition · 156
Encapsulation · 225
end Parameter · 57
even or odd · 185
expanding list · 134
Exponentiation (**) · 102
expressions · 96
extend() · 135
Extending a list · 135

F

falsy · 161
find() · 122
Finding list elements · 138
float() · 98
Floating Point · 65
floor · 102
Floor division (//) · 102
Floor Division (//) · 103
flowchart · 159

for loop · 172
from keyword · 214
function · 53, 187
functions · 183

G

George Boole · 68
Guido van Rossum · 22

I

if condition · 156
immutable · 127
index · 127
index() · 138
Inheritance · 228
input() · 58
Inserting an element to list · 134
installing third party modules · 213
instance · 221
Instance Attributes · 223
int() · 98
Integers · 64
interactive mode · 29
interpreted · 20, 33, 34
interpreter · 75

K

KeyError · 144
keys · 142

L

len() · 115
Length of a Dictionary · 147

Length of a List · 139
Length of a String · 115
library · 212
Library Management System · 233
List · 125, 127
List Concatenation · 140
Logical operator precedence · 90
Logical Operators · 88
Loops · 168
lower() · 120

M

main() function · 247
mappings · 125
Mathematical Expression · 96
mathematical operators · 84
Merging Dictionaries · 147
Methods · 224
Modifying a Dictionary · 145
Modifying a List · 137
modules · 211
Modulo · 105
Modulus (%) · 102
Mu Editor · 23
Multiple Conditions (if-elif) · 157
Multiplication (*) · 99
Mutable · 127

N

Nested Conditionals · 159
Nested Dictionaries · 149
Nested List · 131
NOT · 90
not in · 178
Number of Occurrences · 139
Numbers · 95

Numbers to Strings · 71

O

object · 220
Object-Oriented Programming (OOP) · 220, 225, 228
operand · 84
Operators · 84
OR · 90

P

parameters · 54, 187
pass arguments · 54
PEMDAS · 31
pop() · 136, 146
Popping an Element from a List · 136
position number · 116
Precedence · 101
print() · 55
Printing multiple things · 56
private · 226
Program · 16
Programming Language · 15
project · 233
properties · 222
Python · 19
Python Program · 23

R

random numbers · 106
range() Function · 170
recursion · 191
recursive function · 191
Remove Elements from a List · 135
Removing Items from a Dictionary · 146
Repeating Print · 58

Repetition · 115
REPL Mode · 28
replace() · 122
return keyword · 188
run Python code · 27
running Python · 23

S

Script · 28
Script Mode · 30
sep Parameter · 56
sequence · 170
Sequences · 125
Slicing a List · 132
Slicing Strings · 118
statement · 96
String Creation · 112
String Formatting · 119
String Multiplication · 58
Strings · 66
Strings to Floats · 70
Strings to Integers · 70
strip() · 121
Subtraction (-) · 99
System Architecture · 238

T

Test Scenarios · 238
Third-party modules · 213
title() · 121
Tracing Function Call · 196

Trinket · 26
Triple Quotes · 113
Truthiness · 160
Truthy Vs Falsy · 160
Tuple · 140
Tuple Methods · 142
Tuples · 125
turtle commands · 39
Turtle graphics · 35, 42, 44
type of loop · 172
Types of Modules · 212
Types of Operators · 84

U

uniform() · 107
Unpacking a dictionary · 148
update() · 148
upper() · 120
User Defined Modules · 213

V

values · 47
variable assignment statement · 48
Variable names · 49
Variable Naming Rules · 49
variables · 47

W

while loop · 175

Congrats!

You've reached the end of this book, but your coding journey is just beginning. Every great programmer started where you are now—curious, excited, and ready to explore.

Remember:
- Keep practicing. Build projects, solve puzzles, and challenge yourself.
- Don't be afraid to make mistakes—bugs are just stepping stones to learning.
- Stay curious. There's always more to discover in the world of programming.

A Special Thank You

Thank you for choosing this book as your guide. I hope it has sparked your love for coding and given you the confidence to create amazing things.

Happy Coding!

Made in the USA
Monee, IL
13 March 2025

13957952R00142